DARWIN'S
ON THE ORIGIN OF SPECIES

DARWIN'S
ON THE ORIGIN
OF SPECIES

A MODERN RENDITION

DANIEL DUZDEVICH
WITH A FOREWORD BY OLIVIA JUDSON

INDIANA UNIVERSITY PRESS   *Bloomington & Indianapolis*

This book is a publication of

INDIANA UNIVERSITY PRESS
Office of Scholarly Publishing
Herman B Wells Library 350
1320 East 10th Street
Bloomington, Indiana 47405 USA

iupress.indiana.edu

Telephone orders 800-842-6796
Fax orders 812-855-7931

♾ The paper used in this publication
meets the minimum requirements of
the American National Standard for
Information Sciences–Permanence of
Paper for Printed Library Materials,
ANSI Z39.48 – 1992.

Manufactured in the
United States of America

Library of Congress
Cataloging-in-Publication Data

Library of Congress Cataloging-
in-Publication Data

Duzdevich, Daniel, [date].
  Darwin's On the origin of species : a
modern rendition / by Daniel Duzdevich;
with a foreword by Olivia Judson.
    pages cm
  Includes index.
  ISBN 978-0-253-01166-4 (cl : alk.
paper) – ISBN 978-0-253-01170-1 (pb :
alk. paper) – ISBN 978-0-253-01174-9
(eb) 1. Evolution (Biology) 2. Natural
selection. I. Darwin, Charles, 1809-1882.
On the origin of species. II. Title.
  QH366.2.D89 2014
  576.8'2 – dc23
                          2013034380

1 2 3 4 5   19 18 17 16 15 14

Nagyszüleimnek.

Akik ismerik az élet lényeges dolgait.

CONTENTS

FOREWORD

Olivia Judson

WHY READ *ON THE ORIGIN OF SPECIES* BY CHARLES DARWIN? After all, it was first published more than 150 years ago, and much of the science is out of date. When Darwin was writing, for example, the rules of genetic inheritance had not been figured out, the causes of genetic variation were unknown, and the discovery of the structure of DNA – the molecule that contains genetic information – was almost 100 years in the future. In Darwin's time no one knew how old the earth was, nor that the continents move across its surface. Viruses had not been discovered. In short, the landscape of scientific knowledge was far less filled in.

Nevertheless, there are excellent reasons to read it anyway.

Here's one: It is among the most important books ever written. Publication of the *Origin* transformed biology and transformed our understanding of ourselves. Before the *Origin*, biology was essentially descriptive, an accumulation of unconnected facts and details called "natural history." The patterns in nature were inexplicable. After the *Origin*, the patterns made sense, and the facts and details became part of a vast and sweeping picture. Before the *Origin*, we humans considered ourselves the pet creations of a deity. After the *Origin*, we became part of nature, related to every other being on the planet, with an ancestry that stretches back, back, back, across the eons, to the dawn of life.

It was a huge blow to human vanity. In showing that the earth orbits the sun, not the other way around, Copernicus had removed us from the center of the universe; then Darwin removed our special status as divine creations. Yet seen through the lens of evolution, our myriad imperfections – our capacities for violence and cruelty, our tendencies toward sexual infidelity, our irrationality – become comprehensible. Our finest attributes – our immense capacity for love, kindness, and self-sacrifice; our ability to cooperate; our consciousness; our languages; and our ability to articulate the world and study the universe – become more amazing. With the *Origin* we became one species among hundreds of millions – and the more astonishing for it.

Here's a second reason to read the *Origin:* Darwin was a genius. Thus, as well as being a transformative text, the *Origin* is a window into the mind of one of the world's greatest thinkers. To spend time with such a mind is an inspiration. The *Origin* is a book that, in my experience, becomes more profound on each reading. It is a book you can have a conversation with over a lifetime.

Yet Darwin was not a genius in the traditional mold. He was not obviously brilliant – at university he was an indifferent student – nor was he flamboyant. He did not write beautiful, elegant equations; nor did he wear peculiar costumes, gamble, cut off his ear, or engage in reckless love affairs. Indeed, having concluded that a wife would be "better than a dog," he married his first cousin Emma Wedgwood shortly before his thirtieth birthday and remained with her until his death in 1882; they had ten children, of whom seven survived to adulthood. After his marriage, his life was outwardly dull, consisting of domestic stability, hard work, frequent illness, extensive correspondence, and regular walks. To be sure, there was drama, but it was mostly internal, as he began to realize that his discoveries would wrench human thought in new directions.

Instead of flamboyance and brilliance, Darwin's genius lay in a mix of curiosity and courage, persistence and passion; it lay in great powers of observation and an attention to tiny details. This, after all, was a man who spent eight years writing the definitive study of barnacles, who made important discoveries about orchids, carnivorous plants, and earthworms, and who, as a young man, published a theory of the formation of coral reefs that is still considered correct. More than that – and this is where he is so exceptional – he took this vast accumulation of knowledge and put it all together into a grand new vision of the world. A vision that, in outline and grandeur, remains intact to this day. His work, moreover, can be understood by anyone who cares to try.

Charles Robert Darwin was born on 12 February 1809, the fifth child of a well-to-do family. At the age of sixteen he went to the University of Edinburgh to study medicine but did not stay the course, being horrified by the gory, screaming reality of surgery without anesthesia. When he quit, his father (himself a doctor) told him, "You care for nothing but shooting, dogs, and rat catching, and you will be a disgrace to yourself and all your family." Plan B was to become a clergyman, and so Darwin was dispatched to study at the University of Cambridge. There, by a peculiar coincidence, he lived for a time in rooms that, more than sixty years earlier, had been occupied by William Paley – the natural theologian and moral philosopher who argued that the appearance of design in nature is evidence of a designer (i.e., God), just as a watch is evidence of a watchmaker, and whose writings were part of the curriculum. At Cambridge, Darwin collected beetles, shot birds, read the adventures of the great explorer Alexander von Humboldt, feasted with friends, and otherwise enjoyed himself prodigiously, in an aimless sort of way.

Then things changed. In 1831, when Darwin was twenty-two, he was invited to sail around the world on HMS *Beagle* in the position of companion to the captain – a post for which his family background as a gentleman was one of the chief qualifications. (His father disliked the idea, objecting that it was "a wild scheme" and would make him unfit for being a clergyman. Which it did – though not in the way his father was expecting.) The voyage lasted almost five years and took him to (among other places) Australia, South America, and the Galápagos Islands; it marked the beginning of his intellectual transformation. The *Beagle* returned to England in October 1836, and early in 1837, with all thoughts of becoming a clergyman having been abandoned, Darwin opened his first notebook on the "transmutation of species." That is, on evolution.

Before the *Origin*, most naturalists believed that species were fixed entities created by a deity to fit their particular environments. Darwin himself started out with this belief. Doubt began to seep in while he was on board the *Beagle*, as he observed the living world firsthand and began to notice puzzling facts. If living beings have been specially created for their environments, why do the animals and plants that live on islands resemble those from the nearest mainland, where the environment is often quite different? Why is it possible for an animal "created" for Europe to thrive when introduced to Australia? Some geese almost never swim, yet they have webbed feet. Why? Many fish that live in the total darkness of caves are blind, yet they have the remnants of eyes. Why? Woodpeckers have many traits that make them good at climbing trees and ferreting insects out from under the bark, yet some woodpeckers with those traits eat only fruit, while others live on treeless plains. Why? Why? Why? Such questions perplexed Darwin and gradually led him to ponder evolution.

He was not the first to do so. Toward the end of the eighteenth century, Darwin's grandfather, Erasmus Darwin, published speculations about evolution (in the fashion of the times, many of them were in poetical form). A few years later the great French biologist Jean-Baptiste Lamarck wrote the first major treatise arguing that evolution takes place. Like Darwin, Lamarck started out believing species were immutable and then changed his mind as a result of what he saw in nature. Darwin knew of Lamarck's ideas: during his brief stint as a medical student, he became friends with the comparative anatomist Robert Grant, who spoke of Lamarck enthusiastically.

In 1844 Darwin wrote a long sketch of his own ideas about evolution. But fearing the ruckus it would cause, and feeling that he did not yet have enough evidence to be convincing, he did not make it public. Instead, he told a few friends about it, including the botanist Joseph Hooker, and wrote a letter to Emma, asking her, in the event of his death, to arrange to have it published. He then resumed work he was doing on other subjects.

Someone else was not so timid. Later the same year a Scottish writer, journalist, and publisher named Robert Chambers published – anonymously – a book called *Vestiges of the Natural History of Creation*, in which he put forward a mass of disconnected ideas about evolution, all completely speculative. The book was a sensation. Yet Chambers failed to persuade most people that evolution takes place. Indeed, none of these earlier writers and thinkers had the impact of the *Origin*.

One reason for this is obvious: they all lacked plausible mechanisms by which evolution could occur. Such as natural selection.

I will describe natural selection in a moment. But before I do, consider the fact that humans can, over time, produce huge changes in domestic animals. Just think of the different breeds of sheep, horse, and cow. Or think of the difference between a massive dog like a Great Dane and a tiny one like a Chihuahua. A single Great Dane can weigh more than twenty-five Chihuahuas. Such differences are caused by repeatedly selecting which individuals are allowed to breed with one another. To breed a big dog, then, you repeatedly select the biggest animals of each sex to breed with one another; to breed a little one, you do the reverse. This system works because size can be inherited – that is, it has a genetic component – and because different dogs have different genes. If the breeding is sustained for many generations, the resulting animals will be much larger (or smaller) than their ancestors.

Darwin saw that something similar happens in nature. But rather than humans selecting which traits are desirable, the selection happens as a natural consequence of the fact that different individuals have different genes.

To see what I mean, consider blue tits, small songbirds found throughout Europe and parts of Asia and North Africa. A typical pair has ten chicks in a brood, and some pairs have two broods a year – which is a lot of baby blue tits. So you might expect that the number of blue tits will grow rapidly from one year to the next.

But it does not. Bird census data shows that the number of blue tits does not increase dramatically. What happens to them all? They experience natural selection. That is, they get eaten by cats, or crows, or sparrow hawks – or they escape. They die of hunger, or cold, or disease – or they make it. They fail to find a place to nest, or, having found a nest, fail to attract a mate – or they have a successful mating season. Now, if one bird survives the cold because it is a little fatter or has warmer feathers, or if it escapes from the paw of the cat because it is a little flightier, or if one finds a mate because it is a little sexier, then the fatter, the featherier, the flightier, and the sexier individuals will be more likely to leave offspring. And if the reason for the fatness, featheriness, flightiness, or sexiness has a genetic component, then those traits will start to spread. Conversely, the thin, the unfeathery, the slow, and the unsexy will die – and their genes will be lost. Over time, if the same traits continue to be favored, the characteristics of the whole population will shift.

To put it more brutally, in all populations more individuals are born than can possibly survive and reproduce. Therefore, most individuals die prematurely. But individuals differ from one another in their inherited characteristics – that is, in their genes. If the difference between dying and surviving, between leaving no descendants and leaving some, is in part genetic, then the traits that enhance success – whatever they are – will start to spread, and those that hinder it will start to vanish. This is evolution by natural selection. The ingredients are genetic variation, the death of many coupled with the survival and reproduction of a few, and time.

Moreover, natural selection does not just shape obvious traits, such as how many flowers a plant produces or the number of legs possessed by a starfish. It can shape every aspect of living beings, from complex details of internal physiology to subtle aspects of behavior, as long as there is genetic variation for the traits in question.

Natural selection is the idea most closely associated with Darwin. Yet he was not the first to discover this either. For example, in 1831 – the same year that Darwin embarked on the *Beagle* – a Scottish journalist named Patrick Matthew published a clear and concise account of how natural selection works. Unfortunately for Matthew, however, he included it as an appendix to his book *On Naval Timber and Arboriculture*, and it went unnoticed. Darwin, for one, knew nothing about it until after the *Origin* was published, when Matthew wrote to a magazine to draw it to public attention. (In later years Matthew insisted on putting "Discoverer of the Principle of Natural Selection" on the title pages of his books. This irritated Darwin.)

But there was another discoverer of the principle of natural selection: Alfred Russel Wallace. In the history of biology the story is famous. In February 1858 Wallace wrote to Darwin from the island of Ternate, in what is now Indonesia, where he was collecting birds, beetles, butterflies, and anything else he could catch. The letter contained a manuscript in which Wallace had outlined the idea of evolution by natural selection (though he did not call it that; he referred instead to "a general principle in nature"). He asked Darwin, if he thought the manuscript sufficiently interesting, to forward it to the great geologist Sir Charles Lyell, who also happened to be a close friend and confidante of Darwin. Indeed, Darwin had already revealed his own ideas about evolution by natural selection to Lyell, and Lyell had already urged him to publish them.

On receiving Wallace's letter, Darwin was devastated: he was about to be scooped. But as Wallace had asked, he sent the manuscript to Lyell, along with an anguished letter of his own. ("My dear Lyell... Your words have come true with a vengeance when you said I shd. [sic] be forestalled ... all my originality, whatever it may amount to, will be smashed ... I hope you will approve of Wallace's sketch, that I may tell him what you say.")

The upshot was that Lyell and Hooker arranged for Wallace's paper, along with an excerpt from Darwin's 1844 sketch and a letter that Darwin had previously written to the American botanist Asa Gray, to be read before a meeting of the Linnean Society in London on 1 July 1858. This ensured that both men received credit for the discovery of natural selection. Yet neither man was present at the meeting. Wallace was now in New Guinea, hunting birds of paradise, and Darwin was at home in Kent, mourning the death of his tenth child (from scarlet fever).

The meeting had little obvious impact. In May 1859 the president of the Linnean Society gave a report of the happenings of the previous twelve months. His verdict was that the year "has not, indeed, been marked by any of those striking discoveries which at once revolutionize, so to speak, the department of science on which they bear." Looking back, this seems like one of the great misstatements of all time – but that's because we know what happened next. Darwin, at last, had been galvanized to act. During the next weeks and months he wrote and wrote, drawing together materials, observations, and results amassed over more than twenty years; and late in November 1859, *On the Origin of Species by Means of Natural Selection, or the Preservation of Favoured Races in the Struggle for Life* was published for the first time.

I have one of these first editions in front of me. The binding is green leather; the lettering on the spine is in gold. The volume is thick – just over five hundred pages. When I open it, it smells musty. Which makes me think for a moment of a chilly, misty November day in London – a jostling, Victorian, imperial London, a London of women in crinolines, men in top hats, and children sweeping manure from the streets. Turkey

still had the Ottomans; Russia, the Tsars; convicts were still being transported to Australia. Slaves were picking cotton in the American South.
Back then Germany did not exist as a country; nor did Italy, though it
soon would. The composer Richard Wagner was writing *Tristan und
Isolde*, his great opera of doomed love; Charles Dickens had recently
published *A Tale of Two Cities*. Telephones, cars, airplanes – none existed,
except in imagination.

   And from that earlier world, comes the *Origin*.

The *Origin* is simultaneously an argument and a massive compilation
of evidence, and it is this evidence that, more than anything else, is so
persuasive and gave the book such impact. This is what sets the *Origin*
apart from what was presented to the Linnean Society or from the writings of Patrick Matthew.

   Darwin drew from every branch of biology then known – from fossils, animal embryos, human efforts at plant and animal breeding, the
way animals and plants are distributed around the world, patterns of
extinction, the presence of vestigial organs like the rudimentary eyes of
cavefish, and so on – as well as from observations and numerous experiments, including many he conducted himself.

   To see how his thinking unfolds, consider his treatment of islands.
Darwin points out that the inhabitants of oceanic islands tend to be
similar to – yet clearly different from – the inhabitants of the nearest continent. As an example, he gives the Galápagos Islands. These lie on the
equator more than five hundred miles from the coast of South America,
yet to a great extent the plants and land birds resemble those of the continent. So much so, says Darwin, that a naturalist feels he is "standing
on American land." To explain the resemblances, he proposes that the
animals and plants on islands are not specially created for island life;
rather, they arrive on the islands from the nearest mainland and begin
to evolve in new directions. But if this is right, they need a way to travel
across the ocean. Darwin points out that birds can do this easily – they
can fly – but most land mammals cannot. Which explains why oceanic
islands tend to have many unique species of bird, but (until humans
sailed in with rats, pigs, goats, and other members of their entourage)
no mammals except bats.

But what about plants? How do they travel? At the time, it was "known" that seeds could not survive being soaked in saltwater. But Darwin tested this. He set up jars of saltwater in his study and put seeds into them for various lengths of time; he then removed them, planted them, and investigated whether or not they sprouted. Sure enough, he discovered that many seeds can survive immersion, some for as long as 137 days. He then opened an atlas and, from the rate of the ocean currents, calculated that more than 10 percent of plants have seeds that could float more than nine hundred miles and still grow, should they happen to land in a favorable spot. And that's not the only way plants can get around. They can also be carried by birds. Darwin observes that when birds eat fruit, the seeds often pass through the digestive system intact. A bird blown hundreds of miles off course in a gale could thus carry seeds to far-flung places. More astonishing is that birds that have eaten seeds may in turn be eaten by hawks or owls – and these twice-eaten seeds can *still* germinate. Similarly, if a fish eats a seed and is then eaten by a bird, the bird may become the agent of seed dispersal. This is not mere conjecture. Darwin forced seeds into the bellies of dead fish; fed the dead fish to fishing eagles, storks, and pelicans; and found that some hours later the seeds were either regurgitated or excreted, and that some of these seeds could still grow.

Again and again, Darwin takes an observation – in this case, the patterns of island life – and tests as many implications as he can think of. In doing so, he anticipates and defuses criticism after criticism.

He also makes a number of important and strikingly modern insights. For example, he realized that much of the time the greatest challenges in an organism's environment come from other life forms – predators, parasites, competitors, potential mates, and so on – rather than aspects of the physical environment such as climate. It is, after all, the living environment that generates much of the intricacy in nature – the elaborate displays that have evolved to attract mates, or (in the case of plants) pollinators, the astonishing camouflage with which some animals blend into their surroundings, the various mechanisms by which organisms of all kinds fight off disease. Through evolving interactions, beings such as birds, moths, and flowers can shape each other in complex and beautiful ways.

Moreover, as Darwin recognized, these intricate relationships have broader implications. One is that nature is a web of relationships, and changes that affect life forms in one part of the web can have strong effects elsewhere. In one of his most famous passages he observes that red clover is pollinated only by bumblebees. But the number of bumblebees depends on the number of field mice, because field mice destroy bumblebee nests. The number of field mice is in turn affected by the number of cats, which means that sometimes cats will have a strong effect on the frequency of particular flowers.

A second implication is that evolution is local. So animals and plants living in Australia, or New Zealand, or Mauritius, or wherever, develop an evolving web of relationships among themselves. If an interloper that has evolved in a different web of relationships should happen to arrive, it will sometimes be able to flourish at the expense of the residents. For example, since oceanic islands typically lack ground-dwelling mammals, life forms there repeatedly evolve characteristics that they tend not to evolve if ground-dwelling mammals are present, such as birds losing the power of flight and evolving to nest on the ground. If rats, pigs, or humans should then arrive, these animals find themselves highly vulnerable to predation, or even extinction.

Darwin also realized that although behaviors and mental capacities seem somehow different from traits such as the color of a feather or the scent of a flower, they are not. This means that behaviors can evolve in the same way as any other trait. So if individuals vary in a behavior, and if that variation has some genetic basis, then the behavior can evolve through natural selection.

Over time, then, simple behaviors can potentially evolve into far more elaborate forms. To show how this could happen, Darwin takes the hexagonal combs built by honeybees and asks how the impulse to build such complex structures could have evolved through the slow accumulation of small but useful variations. This passage is one of the finest in the book; he sets up the full difficulty of the problem to be solved and then solves it with great elegance. The problem? Honeybees have evolved to build a complex comb that holds the maximum amount of honey while using the least amount of wax. To solve it, Darwin marshals evidence from other bees that build less complex combs, he writes to a mathematician to investigate the geometry of honeycombs, and he performs experiments to elucidate how honeybees actually build their combs. He concludes that the steps to making complex combs can be reduced to a few simple rules of thumb – and that the ability to follow these rules of thumb can readily evolve if the pressure to economize the use of wax is strong. If, that is, colonies that build more efficient combs tend to perpetuate themselves better than colonies whose combs are less efficient.

But the full magnificence of the *Origin* is this: it presents a worldview of extraordinary power, one that can explain the entire diversity of life on earth. The spiders, wombats, and algae; the earthworms, stag beetles, and luminous fungi; the tigers, oak trees, and all the rest; all the millions of different beings that are here on earth today as well as the countless billions that have lived here before are all bound together into a single tree of life, sculpted by forces of nature that we can identify and understand.

This worldview has several elements: through the tree of life, it gives kinship to all life forms; it predicts that every bacterium, beetle, or breadfruit tree can trace its ancestry back to the origin of life. In principle, then, it's possible to draw a giant family tree, one that includes every being on the planet. But why are there so many different kinds of beings? This is due to what Darwin called "descent with modification," meaning that over time, different populations evolve in different directions. This is, in part, because different populations experience natural selection in different ways: climates can be subtly different, local predators may be different, ditto the food, and so on. It is also, in part, because of what Darwin called "sexual selection," which is the idea that by choosing who to mate with, males and females exert selection upon one another's appearances, songs, scents, and other characteristics. Like natural selection, sexual selection can cause different populations to evolve in different directions. What's more, sexual selection can interact with natural selection. A particularly good example of this interaction comes from recent research on guppies, small fish that live in streams. Female guppies prefer to mate with colorful males. Unfortunately, however, predators often find those males easier to see, so brighter males tend to get killed at a higher rate. Thus, in streams with large numbers of predators, males tend to evolve to be less flamboyant, whereas fewer predators leads to the evolution of more colorful males. Over time – especially immensely long periods of time, such as tens of millions of years – these various processes can produce spectacularly different outcomes in different places.

But there is something else to bear in mind. The substrate for all evolutionary change is genetic variation, which is ultimately caused by mutations to the information contained in DNA. If there is no variation for a trait, then that trait cannot evolve, even if evolution would be advantageous. For example, in one recent experiment, *E. coli* bacteria were grown in an environment that had a limited supply of glucose – the sugar they usually consume – but an abundance of citrate, which they usually cannot consume. It took more than thirty-one thousand generations for a citrate-consuming variant to appear. The reason is that evolving to consume citrate was not a matter of a single mutation, but of several. Still, even for bacteria that go through several generations every day, thirty-one thousand generations is a long time to wait. It also shows that evolution includes a sizeable dollop of chance – in this case, the chance that the necessary genetic variation will appear.

The tree of life, genetic variation, natural selection, sexual selection, chance, and time; these elements are all present in the *Origin*, and the view of evolution it presents is complete in outline. Yet the *Origin* is just that: the beginning of a new way of looking at life on earth, the first word on the subject, not the last. In the decades since 1859, Darwin's ideas have been greatly developed and refined. Darwin himself led the charge; he published five further editions of the *Origin* in his lifetime, often revising heavily from one to the next. Some of the changes are corrections; others, responses to critics; still others incorporate new discoveries. By the sixth edition, which appeared in 1872, the text had grown to include a sketch of evolutionary thinking before the *Origin*, a glossary of terms, and an extra chapter.

But the big breakthroughs came a few decades later, starting with discoveries in genetics. Over the past century, an improved understanding of genetics has generated a far more robust framework for understanding how evolution works. Recent fossil discoveries, as well as better technologies for dating rocks, have vastly improved our understanding of the history of life. This is not surprising; on the contrary, it would be

odd if someone writing more than 150 years ago had understood everything and entirely anticipated modern biology.

When Daniel Duzdevich first mentioned to me that he was putting *On
the Origin of Species* into modern English, my first reaction was *Really?
Do you think it needs it? Chaucer, sure. But Darwin? Is he really that incomprehensible?* On rereading and reflection, I concluded that the answer is
often, yes.

To be sure, parts of the *Origin* are brilliant – lucid, clear, fascinating. Darwin is excellent at description; his best writing is when he is
reporting the results of his experiments, or writing about subjects he
understands well, such as the fossil record. He is sometimes eloquent,
even passionate; the final chapter, where he brings all of his powers of
persuasion to bear, is a triumph of rousing advocacy. But in other places
he is plodding and opaque; the chapter on hybridism, for example, can
defeat the most eager biologist.

He is, unsurprisingly, at his worst when writing about technical subjects he does not understand well or lacks the vocabulary to discuss. The
obvious example is genetics; here, his writing is especially tortured and
hard to follow. But it is easy to see why. He was writing about a subject
that had not been developed yet – the word "gene" was not coined until
early in the twentieth century – and his grasp of how traits are inherited
was foggy. This fog shrouds his prose on the subject. It's a pity he starts
the book with genetics; many readers have surely been discouraged because the opening pages are so turgid.

But perhaps the chief reason that Darwin's prose is difficult is that it
is a kind of high Victorian writing. His sentences are often long, astonishingly convoluted, and sometimes digressive. From a modern stand-
point, they also have peculiar punctuation. Add to that some archaic
turns of phrase, and it's easy to struggle.

Duzdevich is not the first to try to address these problems. In his
book *Almost Like a Whale* (that was the original British title; in the
United States the book was published as *Darwin's Ghost*), geneticist Steve
Jones imagined the text Darwin might have produced if he were writing
now, with a modern knowledge of biology. Jones kept the same chapter

structure and included some of Darwin's original text in facsimile, but the book is nonetheless a radical reinvention of the original. James Costa, an evolutionary biologist, took a more orthodox approach and produced an annotated *Origin*, but while the annotations greatly help the reader to understand the background to Darwin's thinking, they do not make the text any easier. Still another approach has been taken by Mark Ridley, also an evolutionary biologist, with an excellent short book called *How to Read Darwin*. Here, Ridley aims to inspire the reader to persevere. One of his suggestions: don't start at the turgid beginning, but start at chapter 3. And be prepared to make an effort; when Darwin appears to be going on at great length about a subject that seems boring or irrelevant, stop and try to work out why he might be doing so. After all, he was writing for a different audience – an audience with different preconceptions from those we have now. In other words, says Ridley, this is not a book for reading passively, but for tackling and interrogating with gusto.

Such gusto is made much easier by Duzdevich's rendition. For there is no doubt about it: Duzdevich's Darwin is much easier to read than Darwin's Darwin. Indeed, it was this realization that made me enthusiastic about the project and led me to conclude that it's an important contribution, making this fundamental text far more accessible.

So what has Duzdevich done? He has not abridged the text, nor been so crude as to go through it striking out apparently extraneous words. Instead, he has taken the first edition of the *Origin* – which, as the first presentation of a world revolution in human thought, is of the greatest historical interest – and unfolded the sentences and made the syntax and punctuation more modern. He has also replaced some of the more archaic turns of phrase with modern terms. The insights and meaning are all there; the convolution is gone. In other words, he has made a careful translation of the text from Victorian English into twenty-first-century English. The upshot is that the text is much easier to grasp, appreciate, and think about.

Every time I open the *Origin*, I learn something new, and discover further reasons to be impressed by Darwin's breadth of knowledge and depth of understanding. But more than that, the *Origin* and the world-view it contains have transformed the way I think about the planet, the beings on it, and what it means to be human. I find it a source of awe and optimism, reverence and consolation. I hope that, abetted by this new rendition, it will inspire the same feelings in you.

NOTES

I have drawn the details of Darwin's life from Janet Browne's excellent two-volume biography. Volume 1 is called *Charles Darwin: Voyaging*, published by Knopf in 1995. Volume 2 is called *Charles Darwin: The Power of Place*, published by Knopf in 2002. Any reference to the *Origin* is to the first edition.

For Darwin deciding that a wife would be "better than a dog," as well as some other points in favor of marriage, see *Voyaging*, page 379. For his marrying Emma Wedgwood, see *Voyaging*, pages 391–401. For his fleeing the sight of surgery, see *Voyaging*, pages 62–63. For his father's gloomy prediction about Darwin's future, see *Voyaging*, page 89. For Darwin occupying Paley's rooms at Cambridge and for Paley being on the syllabus, see *Voyaging*, page 93. For his lifestyle at Cambridge and his indifference toward his studies, see *Voyaging*, chapters 4 and 5. For details of the offer to travel on the *Beagle* (including a full list of his father's objections to it), see *Voyaging*, chapter 6.

I have drawn Darwin's "puzzling facts" from the pages of the *Origin*.

For Darwin being friends with Robert Grant, and for their discussing Lamarck together, see *Voyaging*, pages 80–83. For Darwin's species sketch of 1844 and some of the reasons he did not want to publish it, see *Voyaging*, pages 445–47. For the impact of *Vestiges*, see *Voyaging*, pages 457–65.

Comparison of the relative weights of Great Danes and Chihuahuas assumes that a Great Dane weighs 174 lbs (78.9 kg) and a Chihuahua, 3 lbs (1.4 kg). For the lives of blue tits, see pages 225–48 of S. Cramp and C. M. Perrins, eds., 1993, *Handbook of the Birds of Europe and the Middle East: The Birds of the Western Palearctic*, Volume 7: *Flycatchers to Shrikes* (Oxford: Oxford University Press). For the parts of the world where you can find them, see page 227; for the number of chicks in a brood, see page 243, column 1. For annual mortality, see page 228.

Patrick Matthew's sketch of natural selection, which appears as the appendix to *On Naval Timber and Arboriculture*, is well worth reading;

the text is available free online from Google Books. For Darwin's irrita-
tion with Matthew, see *The Power of Place*, page 109. For Wallace's letter
and Darwin's reaction to it, see *The Power of Place*, pages 14–17; for Wallace formulating natural selection, see *The Power of Place*, pages 31–33.
For details of how the presentation to the Linnean Society came about
(and Darwin's absence from it), see *The Power of Place*, pages 33–39; for
the meeting itself, see *The Power of Place*, pages 40–41; for the great mis-
statement by the president of the society, see *The Power of Place*, page 42.
All the material presented to the Linnean Society on 1 July 1858 is available at http://wallace-online.org/content/record?itemID=S043.

For Darwin's account of life on oceanic islands, see the *Origin*, chapter 12. For his experiments on seeds, and for his feeding fish stuffed with
seeds to birds, see the *Origin*, chapter 11. His description of bumblebees,
clover, and cats comes from the *Origin*, chapter 3. His description of
honeybees and their combs comes from the *Origin*, chapter 7. For sexual
selection and natural selection in guppies, see J. A. Endler, 1980, "Natural
Selection on Color Patterns in *Poecilia reticulata*," *Evolution* 34: 76–91.
For the experiment on *E. coli*, see Z. D. Blount, C. Z. Borland, and R. E.
Lenski, 2008, "Historical Contingency and the Evolution of a Key Innovation in an Experimental Population of *Escherichia coli*," *Proceedings
of the National Academy of Sciences USA* 105: 7899–7906.

For a full account of the scale of changes between the editions of the
Origin published in Darwin's lifetime, see the introduction to M. Peck-
ham, *The Origin of Species by Charles Darwin: A Variorum Text* (Phila-
delphia: University of Pennsylvania Press, 1959).

Many thanks to Jerry Coyne, Daniel Duzdevich, Dan Haydon,
Gideon Lichfield, Richard Nash, Jean-Olivier Richard, Jonathan Swire,
and especially, Ben Mason for insights, comments, and suggestions.

A NOTE TO THE READER

Daniel Duzdevich

BIOLOGY OF THE VERY SMALL IS WHAT FASCINATES ME. THIS small world is jostled by the motions of water molecules and crisscrossed by intricate chemical reactions. It is a world over which DNA and protein have dominion. And it is a world that unifies life. At the scale of biological molecules, all that is alive proves to be essentially the same. Bacteria swarming in soil, yeast fermenting a lump of moist flour, and humans eating bread are all built from a common cellular machinery. The universalities of molecular biology are explained by a concept central to the *Origin:* all life on earth is related, every species a branch on a single tree. Darwin was the first to recognize this shared ancestry, which we have since discovered to be written into our very molecular makeup. The *Origin* is perhaps even more fascinating today, for all we have learned of the natural world, than it was in 1859.

Evolution is so entangled with the most basic elements of biology that I encountered the subject well before discovering Darwin. The underlying sameness of living things shocked me when I first learned about cells and genes. It still shocks me – the implication that life is continuous and interconnected through time. What struck me the first time I read the *Origin* was that Darwin, through a mass of seemingly disconnected observations, before genetics and before biochemistry, had formulated a theory so powerful that it elegantly accounts for biological universality. It is this insight more than any other – and there are many – that motivated me to pore over Darwin's writing, to try to understand an idea that transformed biology into a science. My interests soon expanded as

I undertook a project to make the *Origin* more accessible to more people. This book is the result.

It is a clear, modern English rendition of the first edition of the *Origin*. It is not an abridgment. The concepts have not been modified or summarized. The sentence and paragraph order have largely been preserved, and the linguistic flavor of the original remains.

A few systematic changes are worth highlighting. Some of Darwin's asides are better suited to footnotes, even though he uses none in the original. In the few cases where I supplement the text with a footnote of my own, they are enclosed by brackets and followed by my initials, as in [This is an editorial footnote. – D.D.].

The word "niche" replaces phrases like "place [of an organism] in the polity/economy of nature." "Niche" had not yet been established as a technical term in Darwin's time, but it is now used in ecology to describe an organism's "place" relative to its environment. "Environment" replaces "conditions of life." Darwin understood that a given environment includes other organisms, so he sometimes specifies just the *physical* conditions of life – that is, the physical environment (climate, for example).

Where Darwin uses "affinity," I generally use "relationship." "Affinity" referred to certain types of similarities among species in the terminology of then-contemporary natural history. In those days, species were grouped according to a variety of schemes, but there was no satisfactory explanation for the features of those groupings. Darwin explained them as a result of common descent so that affinities were transformed into actual relationships between species. He also noticed that similar environments sometimes lead to similar evolutionary solutions to some specific challenge. For example, many cave-dwelling species from wholly distinct branches of the evolutionary tree – such as insects and fish – have evolved blindness. The modern term for this phenomenon is "convergent evolution." Darwin labels it "analogical," or, confusingly, "adaptive"; I use "analogical" for consistency and clarity.

Dissecting Darwin's phrases helps me understand the concepts woven into the *Origin,* concepts that are easy to take for granted in the day-to-day work of the laboratory. Modern molecular biology routinely harnesses the universal qualities of living things. A typical study may involve a bacterial strain coaxed into producing large quantities of a protein native to yeast, a markedly different type of organism. Bacteria and yeast are very distantly related, having diverged from a common ancestor billions of years ago. But that common ancestor already possessed the most fundamental, and the most important, molecular qualities of life – including DNA as the hereditary material, and the machinery needed to manufacture proteins based on information carried by a gene. These characteristics are so important for the very existence of a cell that they cannot be altered in any essential way, so they are passed on from generation to generation, all the way through to the cells growing in a couple of flasks in a laboratory. This continuity explains why bacteria can correctly read a gene from yeast in order to make protein: the two species converse in one molecular language, as all living things do. The phenomenon of shared characteristics applies to successive groupings of the evolutionary tree. All animal embryos, for example, share a common molecular and cellular system for development, inherited from a common ancestor. Darwin observed that such commonalities cluster together in this way, and he recognized their origin: it is common ancestry, which makes the modern science of biology possible, and understandable, and sometimes wonderfully shocking.

ACKNOWLEDGMENTS

MANY PEOPLE CONTRIBUTED TO THE DEVELOPMENT OF THIS project, and I am indebted to all of them. A series of conversations with Walter Bock were formative. Two anonymous reviewers critiqued an early draft and helped me improve the text. Ashley Hennen made insightful and important comments on a later version. The impeccable copyediting work of Jill R. Hughes polished the final draft. Myles Marshall designed the inspired cover and reworked the tree diagram. Thomas Cole created and maintains the companion website. I am especially grateful to Olivia Judson for writing the foreword and for thoughtful discussions. Eric Greene and members of his laboratory were very supportive as I worked to complete the manuscript. Robert Sloan, Nancy Lightfoot, and many others at Indiana University Press made this book a reality.

DARWIN'S
ON THE ORIGIN OF SPECIES

INTRODUCTION

WHEN ON BOARD HMS *BEAGLE* AS NATURALIST, I WAS STRUCK by the distribution of South America's organisms and the geological relationships between its past and present inhabitants. These observations seemed to me to illuminate that mystery of mysteries: the origin of species. After my return home, it occurred to me, in 1837, that this question may be clarified by patiently accumulating and reflecting on all sorts of relevant facts. Following five years' work I began to speculate on the subject and drew up some short notes, which I enlarged in 1844 into a sketch of probable conclusions. Since then I have steadily pursued the same object. I hope the reader will excuse me for entering on these personal details; I just want to show that I have not been hasty in coming to a decision.

My work is nearly finished, but I will require two or three more years to complete it, and as my health is far from strong I have been urged to publish this abstract. I was further prompted to do this because last year Alfred Russel Wallace, who is now studying the natural history of the Malay Archipelago, sent me a memoir in which he arrives at conclusions very similar to mine. He requested that I forward it to Sir Charles Lyell. Lyell and Dr. Hooker, who both knew of my work – Dr. Hooker having read my 1844 sketch – honored me by advising publication of extracts from my own manuscript alongside Mr. Wallace's excellent piece. Both appear in the third volume of the *Journal of the Linnean Society*.

This abstract is necessarily imperfect. I cannot here provide references and must trust the reader's confidence in my accuracy. Errors have no doubt crept in, although I have tried to use only reliable authorities.

I give only general conclusions with a few illustrative examples, which I hope will suffice. And I entirely appreciate the necessity of describing all the information upon which I ground my conclusions in a future work. I am well aware that little is discussed to which additional observations could not be given, often leading to apparently contradictory conclusions. A fair assessment can only be reached by a full statement of the facts and balancing arguments on both sides of each issue; this cannot possibly be done here.

I regret that a lack of space prevents acknowledgment of the generous assistance I have received from many naturalists, some of them personally unknown to me. However, I will not pass up this opportunity to thank Dr. Hooker, who has aided me over the past fifteen years in every possible way with his excellent judgment and large stores of knowledge.

A naturalist considering the origin of species and reflecting on the affinities among organisms, their embryological relationships, their geographic distribution, geological succession, and other factors might conclude that each species had *not* been created independently but had descended as varieties do from other species. But even if such a conclusion were well founded, it would be unsatisfactory until it were shown *how* the innumerable species inhabiting the earth have been modified to a perfection of structure and coadaptation that justly excites our admiration. Naturalists refer to external conditions, such as climate and food supply, as the only possible cause of variation. This may be true in one very limited sense, as we will see, but it's absurd to claim, for example, that the woodpecker's feet, tail, beak, and tongue have become so well adapted to catch insects under tree bark simply due to external conditions. Or consider the mistletoe, which draws nourishment from certain trees, has seeds that must be transported by specific birds, and has flowers with separate sexes requiring particular insects to bring pollen from one flower to the other. It is equally absurd to claim that the structure of this parasitic plant, with its relationships to distinct organisms, results from habits or volition or external conditions. The author of the *Vestiges of the Natural History of Creation* would presumably argue that after an indeterminate number of generations some bird had given birth to a perfectly formed woodpecker and some plant had brought forth a perfectly formed mistletoe. But this assumption is no explanation, leaving the

coadaptations of organisms to one another and their physical environments untouched.

So it is very important to learn how modification and coadaptation happen. As I began my observations, it seemed that a careful study
of domesticated plants and animals would provide the best chance for
insight into this problem. I have not been disappointed; our knowledge
of variation under domestication, though imperfect, invariably affords
the best and safest clue to these and other perplexing problems. I suggest
that this field is highly valuable, though it has commonly been neglected
by naturalists.

I therefore devote the first chapter of this abstract to variation under
domestication. I demonstrate that a large amount of hereditary modification is at least possible and, perhaps more importantly, that humans
have caused huge changes in domesticated plants and animals through
the selection and accumulation of slight successive variations. I then
briefly discuss the variability of species in the wild. This topic could only
have been treated properly by long catalogs of facts, but I nevertheless
discuss the circumstances favorable to variation. The third chapter treats
the struggle for existence among all organisms, which follows inevitably
from their ability to proliferate geometrically: the doctrine of Malthus
applied to all living things. Because many more individuals of each species are born than can possibly survive, any individual possessing even a
slightly favorable variation enjoys a better chance of surviving the complex and sometimes fluctuating environment and is *naturally selected*.
The principle of inheritance ensures that a selected variety will tend to
propagate its new and modified form.

This fundamental subject of natural selection is treated at length
in the fourth chapter, where I discuss how it often causes extinction
of less improved life forms and induces divergence of character. In the
following chapter I address the complex and poorly understood rules
of variation and correlated growth. In the four succeeding chapters I
present the most obvious and serious challenges to the theory: (1) how
a simple organism or simple organ can be changed and perfected into
something highly developed or elaborately constructed, (2) the mental
power of animals (instinct), (3) the infertility of species but fertility
of varieties when crossed (hybridism), and (4) the imperfection of the

geological record. Then I consider the geological succession of organisms through time; in the eleventh and twelfth chapters, their geographic distribution; in the thirteenth, their classification, based on affinities in both embryonic and fully developed states; and in the last chapter I give a brief summary of the work and a few concluding remarks.

Given our ignorance of the relationships among organisms, it is not surprising that much remains to be explained about the origin of species. Who can explain why one species is numerous with a wide range while a related species is rare with a narrow range? And yet such questions are important because their answers explain the present state and future modifications and success of all organisms. We know even less about the relationships among the innumerable past inhabitants of the earth. Although much remains obscure, and will long remain obscure, the most deliberate study and dispassionate judgment of which I am capable have dissuaded me from the view commonly held by naturalists, and previously held by me, that each species has been independently created. I am fully convinced that species are mutable and that species within a genus are linearly descended from some usually extinct species, in the same way that a variety of a given species is a descendant of that species. I am also convinced that natural selection has been the main, though not exclusive, means of modification.

1

VARIATION UNDER DOMESTICATION

IN CONSIDERING THE INDIVIDUALS OF A DOMESTICATED plant or animal variety, it is striking that they are generally more diverse than those belonging to varieties or species in the wild. The vast diversity of domesticated organisms, which have varied under many different climates and treatments, suggests that greater variability results from the conditions under which domestication occurs – conditions unlike those encountered by the parent species in the wild. This variability may partly be connected with excess food, as proposed by Andrew Knight. It seems clear that organisms must be exposed to a new environment over several generations for it to cause appreciable variation, and once organization begins to vary, it usually continues to do so for many generations. There is no case of a variable organism ceasing to be variable under domestication. Established domesticated plants such as wheat still often yield new varieties, and animals domesticated long ago are still capable of rapid improvement or modification.

It is disputed whether the causes of variation – whatever they may be – act during the early or late stage of embryonic development or at the instant of conception. Isidore Geoffroy St. Hilaire's experiments show that unnatural treatment of the embryo causes monstrosities, which cannot be clearly differentiated from mere variations. I strongly suspect that variability is most frequently caused by effects on the egg or sperm before conception, mainly because of the remarkable influence of cultivation or confinement on the functions of the reproductive system, which appear far more susceptible to environmental changes than any other component of organization. Nothing is easier than taming an animal and nothing more difficult than getting it to reproduce in confinement, even when the male and female mate. This is generally attributed to impaired instincts, but many cultivated plants are vigorous yet do not seed. In some cases, minor changes, like a little more or less water at a particular period of growth, determine whether or not a plant will produce seeds. I will not go into the copious details I have collected on this curious subject, but to illustrate the strangeness of the rules that govern the reproduction of captive animals, consider that, with the exception of bears, carnivorous mammals, even from the tropics, breed freely in Britain under confinement, whereas carnivorous birds rarely lay fertile eggs. Many exotic plants have pollen as useless as that of the most sterile hybrids. Some domesticated plants and animals that are otherwise weak and sickly breed freely under confinement; but tame, long-lived, and healthy individuals taken young from the wild may have reproductive systems so seriously affected by unknown causes that they are nonfunctional. Unsurprisingly, then, when the reproductive system actually works under confinement, it does so irregularly, producing offspring that are different from the parents. Finally, some organisms breed under very unnatural conditions – like rabbits and ferrets kept in hutches – demonstrating that their reproductive systems have not been affected. So some organisms withstand domestication and vary only slightly, perhaps hardly more than in the wild.

Sterility is a horticultural nuisance, but variability, the source of all the choicest productions of the garden, shares a cause with sterility. There are many plants (called "sporting" plants by gardeners) that produce single buds or offshoots with novel characteristics, sometimes very different from the rest of the plant. Such buds can be propagated by grafting or other techniques, and sometimes by seed. These "sports" are rare in the wild but common under cultivation. In this case, manipulation of the parent affects a bud or offshoot but not the ovules or pollen. According to most physiologists, however, there is no essential difference between a bud and an ovule in the earliest stages of formation. Therefore, sports show that variability may be largely attributed to the effect on the ovules, pollen, or both by treatment of the parent prior to conception.

In any case, these examples demonstrate that variation is not necessarily connected with the act of generation, as some authors have suggested.

Seedlings from the same fruit and young from the same litter sometimes differ considerably from each other even though both parent and offspring have apparently been exposed to the same conditions, as Müller has remarked. This shows how unimportant direct environmental effects are in comparison to the laws governing reproduction, growth, and inheritance. If the influence of environment were direct, then variation would be the same among offspring. Judging the extent to which heat, moisture, light, food, and other factors have an impact on variation is difficult. My impression is that such agents produce very little direct effect on animals, but apparently more on plants. (Mr. Buckman's recent experiments on plants are valuable here.) When all or nearly all individuals exposed to certain conditions are identically affected, the resultant changes *appear* to flow directly from the conditions. But in some cases opposite conditions generate similar structural changes. Nevertheless, some slight amount of change may be attributed to direct environmental action, as in certain cases of increased size from greater food intake, altered coloration from particular kinds of food and light, and perhaps the thickness of fur from climate.

Habit also has a deciding influence, as with the flowering period of plants transported from one climate to another. The effect is greater in animals. For example, I find that the wing bones of a domestic duck weigh less and the leg bones weigh more in proportion to the whole skeleton than do those of a wild duck. I presume this results from the domestic duck flying much less and walking more than its wild parent. The large inherited udders of cows and goats in countries where they are habitually milked is another example of the effect of use. There is no domestic animal that in some region does not have drooping ears. As suggested by some authors, this probably results from the disuse of ear muscles, the animals being rarely alarmed.

Many rules regulate variation; some of them can be dimly seen and will be briefly mentioned. Here I will only allude to "correlated growth." For example, a change in the embryo or larva often entails changes in the mature animal. With monstrosities, correlations between distinct parts are very curious.[1] Breeders maintain that long limbs are often accompanied by an elongated head. Some correlations are whimsical; for example, cats with blue eyes are invariably deaf. There are many remarkable cases among plants and animals of coloration and constitutional peculiarities going together. Observations collected by Heusinger suggest that white sheep and pigs are affected differently by poisonous vegetables than individuals with coloration. Hairless dogs have imperfect teeth; long-haired and coarse-haired animals tend to have long or many horns; pigeons with feathered feet have skin between their outer toes; pigeons with short beaks have small feet, and those with long beaks have large feet. If humans select and augment a peculiarity, they will probably unintentionally modify other parts as a consequence of the mysterious rules of correlated growth.

The dimly seen or unknown rules of variation yield infinitely complex and diverse results. Treatises on some established domesticated plants, such as the hyacinth, potato, and even the dahlia, reveal a surprising number of slight structural and constitutional differences between varieties and subvarieties. The whole organization seems to have become plastic, with a tendency to depart somewhat from the parental type.

Variations that cannot be inherited are unimportant to this argument. What nevertheless remains is an endless number of diverse *heritable* structural deviations of both slight and considerable physiological importance. (Dr. Prosper Lucas's two-volume work is the best treatment of this subject.) The strong propensity for inheritance is known by breeders, whose fundamental belief is that "like produces like." (Only theoretical writers have thrown doubt on this principle.) When a commonly occurring deviation is observed in both parent and offspring, it may result from the same cause acting on both. But when a very rare deviation due to some extraordinary combination of circumstances appears, say, once in several million individuals all apparently exposed to the same conditions, and it reappears in an offspring, the mere doctrine of chance almost compels us to attribute its reappearance to inheritance.

1. St. Hilaire gives many examples in his great work on this subject, *Histories des Anomalies.*

Everyone has heard of albinism, prickly skin, hairy bodies, and other such peculiar characteristics reappearing in several members of the same family. If strange and rare deviations really are heritable, then surely commonplace deviations are also heritable. Perhaps the correct view is to take inheritance of every characteristic as the rule and non-inheritance as the anomaly.

The laws of inheritance are unknown. It is unknown why some given peculiarity of individuals within the same species, or of individuals among different species, is sometimes inherited and sometimes not, why a child reverts to characteristics found in a grandparent or more remote ancestor, or why some peculiarities are inherited in a gender-dependent manner.[2] Peculiarities appearing in the males of domestic breeds are often transmitted exclusively, or more strongly, to male progeny. A more important rule is that the age at which a peculiarity first appears tends to be the same in the parent and in its offspring (although sometimes earlier in the offspring). In many cases this cannot be otherwise. For example, the inherited peculiarities of cattle horns can appear only as the offspring mature, and peculiarities in the silkworm are known to appear at the corresponding caterpillar or cocoon stage. Hereditary diseases and some other examples suggest that this rule is generally applicable: even when there is no apparent reason for a peculiarity to appear at a particular stage, it tends to appear in the offspring at the same period of development as in the parent. This is very important to illuminating the rules of embryology. These remarks are, of course, confined to the first *appearance* of a peculiarity and not to its primary *cause*, which may have acted on the egg or sperm. If the offspring of a short-horned cow and a long-horned bull develops long horns, then it's clearly due to the sperm.

Naturalists often argue that when domestic varieties run wild, their characteristics gradually but surely revert to those found in the original stocks, and that, consequently, deductions drawn from domestic varieties cannot be applied to species in nature. I have tried without success to find the decisive facts on which this statement is so often and so boldly

2. [These phenomena are now well understood through genetics. – D.D.]

made; it would be very difficult to prove, because many established domestic varieties could not possibly survive in the wild. In many cases we
do not know what the original stock was and could not tell whether or
not reversion had ensued. It would also be necessary to turn loose only
one variety to avoid the effects of intercrossing. Nevertheless, varieties
sometimes do partially revert to the parental form. For example, if various strains of cabbage were cultivated in very poor soil for many generations, they would probably revert wholly or largely to the wild stock.
(However, some effect would have to be attributed to the direct action
of the poor soil.) Whether or not the experiment would succeed is not
particularly important to the argument, because the experiment necessarily alters the environment. If a strong tendency for reversion – that is,
a loss of acquired characteristics under constant conditions in a large
population so that free crossing, by blending, checks slight deviations of
structure – were demonstrated in domesticated varieties, I would grant
that nothing deduced from domestic varieties would apply to species.
But there is not a shadow of evidence in favor of this view. To assert that
we could not breed cart and racehorses, long- and short-haired cattle, and
poultry of various breeds, and cultivate edible vegetables for an almost
infinite number of generations is contrary to all experience. When the
environment changes in nature, variations and reversions probably *do*
occur, but natural selection, as will be explained, determines how far
such new characteristics are preserved.

As already mentioned, there is *less* uniformity of character among
individuals of a domestic variety than among individuals of a true species. Also, domestic varieties of the same species often have a monstrous
character, by which I mean that although they differ in some minor respects from one another and members of the same genus, they often
differ extremely in some one part. With these exceptions and that of the
perfect fertility of crossed varieties (discussed later), domestic varieties
of the same species differ from one another in a manner similar to the
way closely related species of the same genus differ in the wild. There are
very few domestic varieties of plant or animal that have not been classified by some competent judges as just varieties and by others as descendants of distinct parent species; if there were any significant distinction
between domestic varieties and species, this source of doubt would be

less common. Contrary to frequently made assertions, I think domestic varieties differ from one another in generic characteristics,[3] which naturalists disagree in defining because all such valuations are currently empirical. Given the following examination of the origin of genera, there is no reason to often expect generic differences in domesticated organisms.

Attempts to estimate the amount of structural difference between domestic varieties of the same species are hampered by our ignorance of whether they have descended from one parent species or several; it would be interesting to clear up this problem. For example, if it were shown that the greyhound, bloodhound, terrier, spaniel, and bulldog, which propagate their kind truly, are derived from a single species, the supposed immutability of the many closely related natural species (such as the foxes) would be brought under considerable doubt. I do not believe that all dog breeds have descended from one wild species (see below),[4] but there is tentative or even strong evidence that some other domestic varieties have.

Humans are often assumed to have chosen for domestication those plants and animals that possess an extraordinary inherent tendency to vary and to withstand diverse climates. Although such capacities have added significantly to the value of many domesticated productions, how could primitive humans have possibly known when first taming an animal that it would vary in succeeding generations and endure other climates? The limited variability of the ass and the guinea fowl, and the low tolerance for warmth by the reindeer and for cold by the common camel did not prevent their domestication. If plants and animals equal in number and belonging to equally diverse classes and regions to existing domesticated organisms were taken from the wild and bred for an equal number of generations under domestication, they would vary on average as much as the parent species of already domesticated organisms have varied.

3. ["Generic characteristics" refer to those that are relevant at the genus level. – D.D.]
4. [In fact, they have; all dogs belong to the same species, descended from the wolf. – D.D.]

I think it is impossible to ascertain with complete certainty whether established domesticated plants and animals have descended from one or multiple species. Those who believe in the multiple origin of domestic animals argue mainly that ancient records, especially on the monuments of Egypt, reveal a great diversity of breeds, some of which resemble or are identical to existing ones. Even if this were found to be more strictly and generally true than I believe is the case, it suggests only that some of our breeds originated there four or five thousand years ago. Based on Mr. Horner's research, civilization advanced enough to manufacture pottery probably existed in the Nile valley thirteen or fourteen thousand years ago; it is not known how long before these ancient periods peoples like those of Tierra del Fuego or Australia, who possess a semi-domesticated dog, may have existed in Egypt.

I think the whole subject must remain vague. Nevertheless, without going into details – but based on geographic and other considerations – I think it is likely that domestic dogs have descended from several wild species. I cannot form an opinion with respect to goats and sheep. Information about the habits, voice, constitution, and other features of humped Indian cattle, communicated to me by Mr. Blyth, indicate that it descended from a different stock than European cattle, which, in turn, have more than one parent, according to several judges. And for reasons I cannot cover here, I am doubtfully inclined to believe, in opposition to several authors, that all the varieties of horse have descended from one wild stock. Mr. Blyth – whose opinion I value highly, drawn as it is from his large and varied stores of knowledge – thinks that all poultry breeds have proceeded from the common wild Indian fowl. Duck and rabbit breeds, which differ considerably from one another in structure, have all descended from the common wild duck and rabbit.

Some authors carry the doctrine of plural descent to an absurd extreme, believing that *every* variety that breeds true, even those possessing only very slight distinctive characteristics, has a distinct wild prototype. At this rate there must have existed twenty species of wild cattle and sheep – and several goats – in Europe alone, and even several within Great Britain. One author believes that there once existed eleven unique wild sheep species in Great Britain! Britain barely has any unique mammals, France only a few distinct from Germany, and vice versa, and the same is true of Hungary, Spain, and other localities. Yet each of these kingdoms possesses several peculiar breeds of cattle, sheep, and other animals, meaning that many domestic breeds originated in Europe because it lacks sufficient unique species as parent stocks. This is also true for India. Even in the case of the domestic dogs of the whole world, which I admit probably descended from multiple wild species, there has been immense heritable variation. Who can believe that animals closely resembling the Italian greyhound, bloodhound, bulldog, or Blenheim spaniel – so unlike all wild dogs – ever existed in the wild? Crossing a few original dog species would result in only intermediate forms, and accounting for domestic dogs by this process requires the previous exis-tence of extreme forms, similar to the breeds mentioned, in a wild state. In addition, the possibility of generating distinct varieties by crossing is exaggerated. Obviously a variety can be modified by occasional cross-ing if mongrels with desired characteristics are carefully selected, but a variety intermediate between two extremely different varieties or spe-cies could not be obtained. (Sir J. Sebright experimentally attempted precisely this and failed.) The offspring from the first cross between two pure breeds is fairly uniform – and as I have found with pigeons, some-times very uniform – and everything seems simple enough. But when these mongrels are crossed with one another for several generations, few will be alike and the utter hopelessness of the task becomes apparent. An intermediate between two very distinct breeds could be obtained only with extreme care and continuous long-term selection, but I cannot find a single recorded case of a variety formed this way.

Believing that it is best to study one particular group, after deliberation I took up domestic pigeons. Pigeons have been watched, tended with the utmost care, and loved by many people. They have been domesticated for thousands of years in several parts of the world. Professor Lepsius has pointed out to me that the earliest known record of pigeons is in the fifth Egyptian dynasty (ca. 3000 BC), but Mr. Birch informs me that pigeons are given a bill of fare in the previous dynasty. In Roman times pigeons fetched a high price. Pliny writes, "Nay, they are come to this pass, that they can reckon up their pedigree and race." The court of India's Akber Khan (ca. 1600) always traveled with at least twenty thousand pigeons.

"The monarchs of Iran and Turan sent him some very rare birds," and, continues the court historian, "His Majesty, by crossing the breeds, which method was never practiced before, has improved them astonishingly." At about this same period, the Dutch were as enthusiastic about pigeons as the Romans had been.

I have kept every pigeon breed that I could purchase or obtain and have been kindly favored with skins from several parts of the world, especially by the Hon. W. Elliot from India and the Hon. C. Murray from Persia. Many treatises in several languages have been written about pigeons, some of them very important because of their age. I have consulted several eminent breeders and joined two of the London Pigeon Clubs. The diversity of breeds is astonishing. Compare the English carrier to the short-faced tumbler and see the wonderful differences in their beaks, with corresponding differences in their skulls. The carrier, especially the male, is remarkable for the carunculated skin about its head and the accompanying greatly elongated eyelids, large nostrils, and widely gaping mouth. The outline of the short-faced tumbler's beak is like that of the finch's beak, while the common tumbler has the strictly inherited singular habit of flying at a great height in a compact flock and tumbling in the air head-over-heels. The runt is large with a long, massive beak and big feet. Some runt sub-breeds have long necks; others, long wings and tails; still others, short tails. The barb is related to the carrier but has a short and broad beak instead of a long one. The pouter's body, wings, and legs are elongated; it glories in inflating its enormously developed crop, which may elicit astonishment and even laughter. The turbit has a short and conical beak, with a line of reversed feathers down its breast and a habit of slightly expanding the upper part of its esophagus. The Jacobin has feathers along the back of the neck that are so reversed that they form a hood; for its size, it has relatively elongated wing and tail feathers. The trumpeter and laugher, as their names suggest, coo very differently from the other breeds. The fantail has thirty or even forty tail feathers – even though the normal number in all members of the pigeon family is twelve or fourteen – and they are kept expanded and carried so erect that in good specimens the head and tail touch; the oil gland is aborted. Several other less distinct breeds could be mentioned.

Skeletal structure also differs among breeds. The length, breadth, and curvature of face bones differ enormously. The size and shape of the ramus of the lower jaw, the size and shape of the apertures in the sternum, and the degree of divergence and relative sizes of the two arms of the furcula vary remarkably. The number of caudal and sacral vertebrae; the number of ribs, along with their relative breadth and presence of processes; the proportional width of the mouth; the proportional length of the eyelids, nostrils, tongue – not always in strict correlation with the length of the beak – the size of the crop and the upper esophagus; the development and abortion of the oil gland; the number of primary wing and caudal feathers; the relative lengths of the wings and the tail to each other and to the body; the relative lengths of the legs and feet; the number of scutellae on the toes; and the development of skin between the toes are all structural features that vary. The period at which complete plumage is acquired, the state of the down with which nestlings are clothed when hatched, and the size and shape of the eggs also vary. Flying style and, in some breeds, voice and disposition differ remarkably. Lastly, in certain breeds the males and females have come to differ slightly from each other.

There are at least twenty pigeon breeds that an ornithologist would classify as well-defined species if he were told they were wild birds. I doubt any ornithologist would place the English carrier, short-faced tumbler, runt, barb, pouter, and fantail in the same genus, especially because for each there are several true-breeding sub-breeds, or, as he would call them, species.

Despite the great differences among pigeon breeds, I agree with the common opinion of naturalists that they have all descended from the rock pigeon (*Columba livia*), a category that includes several geographical varieties or sub-species differing from one another in minor respects. Some of the justifications for this conclusion are somewhat applicable in other cases, so I will briefly give them here. If pigeon breeds are *not* varieties and have *not* descended from the rock pigeon, then they must have descended from at least seven or eight original stocks, because it would be impossible to generate the present domestic breeds by the crossing of any fewer. For example, how could a pouter be produced by crossing unless one of the parental stocks possessed the characteristically enormous crop? The supposed original stocks must all have been *rock* pigeons – that is, not breeding or perching in trees. But besides *C. livia* only two or three other species of rock pigeon are known, and these lack the characteristics of domestic breeds. Therefore, the supposed original stocks would have to either (1) still exist unknown to ornithologists in the regions where they were first domesticated, or (2) have become extinct. Because of the rock pigeon's size, habits, and remarkable characteristics, (1) is unlikely. And birds that are good fliers and breed on precipices are unlikely to become extinct. The common rock pigeon has the same habits as the domestic breeds and hasn't been exterminated on several British islets or the shores of the Mediterranean, making (2) a rash assumption. Furthermore, the above-mentioned breeds have been introduced to all parts of the world, so some of them must have wound up in their supposed native regions, yet not one has ever became wild or feral (although the dovecot pigeon, which is the rock pigeon in a slightly altered state, has become feral in several places). All recent experience demonstrates the difficulty of breeding wild animals in confinement, but the hypothesis of plural origin for domestic pigeons implies that at least seven or eight species were so thoroughly domesticated in ancient times by half-civilized humans that they were quite prolific under confinement.

An important argument, also relevant to other cases, is that although these breeds are generally the same in constitution, habits, voice, coloring, and most structural parts as the wild rock pigeon, they are extraordinarily abnormal in other parts of structure. We could search the entire pigeon family in vain for a beak like the English carrier's, short-faced tumbler's, or barb's; for reversed feathers like the Jacobin's; for a crop like the pouter's; or for tail feathers like the fantail's. So it would have to be assumed not only that half-civilized humans thoroughly domesticated several species but also that they intentionally or unintentionally picked abnormal species, and that these species are now all unknown or extinct. So many strange contingencies seem very improbable.

Coloration is worth considering. The rock pigeon is slate blue with a white rump (the Indian sub-species, Strickland's *C. intermedia*, has a bluish rump). The tail has a terminal dark bar with the bases of the outer feathers edged in white; the wings have two black bars; some semidomestic and apparently wild breeds also have black checkers on the wings. These marks do not occur together in any other species of the whole family. *All of these marks, down to the white edging of the tail feathers, sometimes develop perfectly in thoroughly well-bred specimens from every one of the domestic breeds.* And when two birds from distinct breeds lacking these marks and blue coloring are crossed, the mongrel offspring are very likely to acquire them. I crossed some white fantails with black barbs and got mottled brown and black offspring. Then I crossed these together and one of the offspring had as beautiful a blue color, white rump, black double wing bar, and barred white-edged tail feathers as any wild rock pigeon! These observations can be understood through the well-known principle of reversion to ancestral characteristics, if all domestic breeds have descended from the rock pigeon. The alternative would require making one of two very unlikely assumptions: (1) all of the supposed multiple original stocks were colored and marked like the rock pigeon, even though no such species exists today, so that each breed reverts to the same characteristics, or (2) even the purest breed has been crossed by the rock pigeon within twelve or twenty generations.[5]

Hybrid offspring from between all domestic pigeons are fertile. I can state this from my own crosses, intentionally made between the most distinct breeds. It is difficult, if not impossible, to suggest a single case of fertile hybrid offspring from two unambiguously distinct *species*. Some authors believe that long-term domestication eliminates this strong tendency for sterility. This hypothesis may be true if applied to closely related species, based on the history of the dog, but is unsupported by a single experiment. However, it would be rash to extend the hypothesis and claim that supposedly original "species" as distinct as the carrier, tumbler, pouter, and fantail could have produced fertile offspring when crossed.

5. There is no support for the idea that offspring revert to ancestral characteristics introduced more than twenty generations ago. In a breed that is crossed with another distinct breed only once, the tendency for reversion obviously diminishes with each generation as the foreign contribution thins. But if parents revert to a characteristic lost by some former generation even in the absence of crosses with a distinct breed, then the tendency to revert can be transmitted undiminished and indefinitely. These two cases are often confused in treatises on inheritance.

I feel no doubt that all domestic pigeon breeds have descended from *Columba livia* and its geographical sub-species. To reiterate, the reasons are: (1) it is unlikely that primitive humans got seven or eight supposed pigeon species to breed under domestication, with none of these supposed species existing today and none of the breeds having become feral in their supposed native regions; (2) these species have certain abnormal characteristics with respect to the whole pigeon family but are like rock pigeons in other respects; (3) the blue color and marks of the rock pigeon occasionally appear in all breeds both when kept pure and when crossed; and (4) mongrel offspring are fertile.

There is even further support for my assertion. The rock pigeon has been domesticated recently in Europe and India, agreeing in habit and many structural characteristics with all domestic breeds. Furthermore, it is possible to make an almost perfect incremental series between extremes of structure using sub-breeds within any one breed, especially if we include specimens from distant regions. Also, the main distinctive feature of each breed is highly variable. These considerations will be invoked in discussing selection as explaining the immense amount of variation pigeons have undergone. The reason that the breeds often have such monstrous characteristics will also be explained.

When I first kept pigeons, I felt as much difficulty in believing that they have descended from one parent as any naturalist would about the many species of finches or other large bird groups in the wild. It was striking to me that every breeder of domestic animals and every cultivator of plants with whom I talked or whose treatises I read is convinced that each breed has descended from a distinct original species. A celebrated breeder of Hereford cattle would laugh with scorn at the suggestion that his livestock have descended from long-horns. I have never met a pigeon, poultry, duck, or rabbit breeder who was not fully convinced that each main breed has descended from a distinct species. In his treatise on pears and apples, Van Mons rejects that the several varieties (such as Ribston pippin and Codlin apple) could ever have proceeded from seeds of the same tree. There are innumerable other examples. The explanation, I think, is simple: long-term study impresses on the mind *differences* between breeds, and although they know that individuals of each breed vary slightly – prizes are won by the selection of such slight

differences – they fail to sum up in their minds how, over many generations, slight differences can accumulate into large differences. There are
naturalists who know less about inheritance and no more about intermediate links in the lines of descent than breeders but nevertheless admit
that many domestic varieties have descended from common parents. Yet
they deride the idea of species in nature being lineal descendants of other
species. Perhaps they should be more cautious.

What are the steps by which a domestic variety arises from one or several
related species? Environmental conditions and habit may play a minor
role, but they cannot account for the differences between a dray and a
racehorse, a greyhound and a bloodhound, or a carrier and a tumbler
pigeon. One of the most remarkable features of domesticated organisms
is that we see in them adaptations, but to human use or fancy rather than
the animal or plant's own good. Some of these useful variations probably
appeared suddenly. According to many botanists, the fuller's teazle, with
its hooks unrivaled by any mechanical device, is a variety of the wild *Dip-
sacus* that arose suddenly in a seedling. The same is probably true of the
turnspit dog and known to be true of the ancon sheep. But on comparing the dray horse and racehorse; the dromedary and camel; the various
sheep breeds fit for cultivated land or mountainous pasture and each
with wool for a different purpose; the various dog breeds each uniquely
useful to humans; the gamecock (so pertinacious in battle) with breeds
that are not quarrelsome, with "everlasting layers" (which never sit), with
the bantam (so small and elegant); and the host of agricultural, culinary,
orchard, and flower-garden breeds of plants useful to humans at different seasons and for different purposes, or so beautiful in their eyes, mere
variability does not suffice. Every breed could not have been suddenly
produced perfectly useful; in some instances this is historically known
not to be the case. The key is the human power of cumulative selection:
nature provides successive variations and humans add them up in useful
directions, thus producing different breeds.

   The great power of selection is not hypothetical. Several eminent
breeders have drastically modified some sheep and cattle breeds even
within a single lifetime. To fully appreciate what they have done, it is
necessary to read some of the many treatises on the subject and actually

inspect the animals. Breeders habitually speak of animals' organization as plastic, something they can mold almost as they please. (If I had space I could quote numerous passages to this effect from expert authorities.) Youatt was a good judge of animals and probably more knowledgeable about the work of agriculturalists than anyone; he describes selection as "that which enables the agriculturalist, not only to modify the character of his flock, but to change it altogether. It is the magician's wand, by means of which he may summon into life whatever form and mold he pleases." Speaking of breeders' feats with sheep, Lord Somerville says, "It would seem as if they had chalked out upon a wall a form perfect in itself, and then had given it existence." The skillful breeder Sir John Sebright used to say about pigeons that "he would produce any given feather in three years, but it would take him six years to obtain head and beak." The importance of selection in breeding merino sheep is so fully recognized in Saxony that the use of the principle has become a trade. The sheep are placed on a table and studied like a picture by a connoisseur three times over the course of several months. The sheep are marked and classed so that the very best can ultimately be selected for breeding.

What English breeders have effected is proven by the high prices given for animals with a good pedigree, which have been exported almost everywhere in the world. The improvement is not generally a result of crossing different breeds – a practice opposed by the best breeders, with the exception of occasional crosses among closely related sub-breeds. When a cross *is* made, careful selection is even more important than usual. If selection involved simply separating some very distinct variety for breeding, then it would be too obvious to even discuss. But its importance lies in the dramatic effect of unidirectionally accumulating minute differences imperceptible to the untrained eye – mine included – over many generations. Very few men are discerning enough and in possession of the proper judgment to become eminent breeders. Even with these gifts, he will only make improvements by studying his subject for years and devoting a lifetime to his task with indomitable perseverance. Few would believe the talent and practice necessary to become even a skillful pigeon breeder.

These concepts also apply to horticulture, although the variations tend to be more abrupt. No one supposes that the choicest productions arose by a single variation in an original stock. This can be proven in some cases, for records have been kept; the steadily increasing size of the gooseberry is one example. The astonishing improvements in florists' flowers are apparent when present-day varieties are compared with drawings made only twenty or thirty years ago. Seed raisers do not need to pick out the best specimens from a well-established plant variety, but simply pull up the "rogues," as they call plants that deviate from the defined standard. This kind of selection is also followed with animals, because no one is careless enough to let the worst animals breed.

Another way to observe the accumulated effects of selection is by comparing the diversity of flowers of different varieties of one flower-garden species; the diversity of leaves, pods, tubers, or other valued parts of kitchen-garden species; or the diversity of fruits of a species in the orchard relative to other parts of the same variety. Notice how different the leaves of a cabbage are but how alike the flowers; how different the flowers of a heartsease are but how alike the leaves; how much the gooseberry fruit differs in size, color, shape, and fuzziness but how similar the flowers are. It's not that varieties differing drastically in one way don't differ in others. In fact, this is rarely the case. Correlated growth, which should never be underestimated, ensures some differences in other parts. However, as a general rule, selection for slight variations in a specific part will produce varieties differing mostly in that part.

Some may object that the principle of selection has been reduced to methodical practice for less than seventy-five years. It is true that it has been utilized frequently in recent years and accompanied by rapid and important results with many treatises published on the subject. All the same, it is *not* a modern discovery. I could give several references that acknowledge its importance in works of high antiquity. In rude and barbarous periods of English history, choice animals were often imported and laws were passed to prevent their exportation. The destruction of horses under a certain size was ordered, comparable to the "roguing" of plants mentioned earlier. I found that the principle of selection is clearly given in an ancient Chinese encyclopedia. Some of its explicit rules are written down by Roman classical writers. It is clear from passages in Genesis that the color of domestic animals was attended to at that early period. Indigenous groups sometimes cross their dogs with wild canines to improve the breed, both in the present day and – as attested by Pliny – in the past. The natives of South Africa mate their draft cattle by color, just as some of the Eskimo do for their dog teams. Livingstone reports that good domestic breeds are valued by inhabitants of Africa's interior who have not associated with Europeans. Some of these examples do not demonstrate actual selection, but they show that breeding of domestic animals was done in ancient times and is now done by indigenous peoples. It would be strange if attention had *not* been paid to breeding, the inheritance of good and bad qualities being so obvious.

Today, breeders try to make a superior and novel strain or sub-breed by methodical selection with a preconceived object in view, but for this discussion a type of selection I call "unconscious" is more important; it results from everyone trying to possess and breed from the best individual animals. For example, a man who keeps pointers will obviously try to obtain the best dogs he can and then breed from the best of his own. He has no intention or expectation of permanently altering the breed, but this process, extended over centuries, will modify any breed. Using this same process, only more methodically, Bakewell, Collins, and others greatly modified, even during their own lifetimes, the forms and qualities of their cattle. These kinds of gradual and unobservable changes could never be recognized in the absence of measurements or careful drawings made long ago for comparison. In some cases, unchanged or slightly changed individuals of a known breed can be found in less civilized regions where the breed has been less improved. The King Charles spaniel may have been extensively modified unconsciously since the time of its namesake. Some authorities assert that the setter is directly derived from the spaniel, and the English pointer is known to have been greatly changed over the last century, probably by crosses with the foxhound. Importantly, change has been effected unconsciously and gradually but so effectively that although the old Spanish pointer came from Spain, there is no native dog in Spain like the English pointer.[6]

Through a similar process of selection, coupled with careful training, English racehorses have come to surpass their parent Arab stock in speed and size so that by the regulations of the Goodwood Races,

6. I am told by Mr. Barrow, who has not observed any.

they are favored in the weights they carry. Lord Spencer and others have shown that English cattle have increased in weight and early maturity compared to the stock once kept in this country. The stages through which carrier and tumbler pigeons have passed, and how they have come to differ so greatly from the rock pigeon, can be traced by comparing accounts given in old treatises with modern British, Indian, and Persian breeds.

Youatt gives an excellent example of unconscious selection in which the breeders in question could never have expected or even wanted to produce two distinct strains. He remarks that the two flocks of Leicester sheep kept by Mr. Buckley and Mr. Burgess "have been purely bred from the original stock of Mr. Bakewell for upward of fifty years. There is not a suspicion existing in the mind of any one at all acquainted with the subject that the owner of either of them has deviated in any one instance from the pure blood of Mr. Bakewell's flock, and yet the difference between the sheep possessed by these two gentlemen is so great that they have the appearance of being quite different varieties."

There may be peoples so barbarous that they never consider the inherited characteristics of their domestic animals, but if an animal is particularly useful to them for some special reason, it will be preserved during famines and other accidents and consequently leave more offspring than its inferior brethren; this is a kind of unconscious selection. The value of animals is demonstrated even among the barbarians of Tierra del Fuego who kill and devour their old women during times of dearth, as of less value than their dogs.

In plants the same gradual process of improvement through occasional preservation of the best individuals can be recognized in the larger size and more intense beauty of modern heartsease, rose, geranium, dahlia, and other varieties when compared to older varieties or parent stocks. (This applies whether or not the individuals can be classified as belonging to a distinct variety on first appearance, and whether or not species or varieties had been blended by crossing.) No one would expect to get a first-rate heartsease, dahlia, or melting pear from the seed of a wild plant. Although the pear was cultivated in classical times, Pliny's descriptions suggest it was a fruit of very inferior quality. Horticultural essays convey surprise about the gardener's wonderful skill in generating splendid products from such poor materials, but the art has been simple: the final result proceeds from an almost unconscious process. It has always involved cultivating the best-known variety, sowing its seeds, and selecting slightly better varieties when they happen to appear. The gardeners of classical times cultivated the best pear they could procure with no intention of providing us with such sweet fruit, and yet in part we owe to them our pear, because they naturally chose and preserved the best varieties they found.

The large amount of change that has been slowly accumulated unconsciously in cultivated plants explains why we cannot recognize the parent stocks of many established kitchen- and flower-garden varieties. If it took centuries or millennia of improvement to create useful plants, we can understand why Australia, the Cape of Good Hope, and other regions inhabited by uncivilized peoples have not afforded us a single plant worth culture. It is not that these species-rich countries lack original stocks of useful plants, but that these native plants have not been improved to a standard of perfection by continuous selection as in anciently civilized regions.

Domestic animals kept by uncivilized peoples almost always have to struggle for their own food, at least during certain seasons. Individuals of a single species with slightly different constitutions or structure often succeed better in one environment than another; by a process of "natural selection," as will be explained, two sub-breeds might form. This may partly explain the observation of some authors that varieties kept by natives are more like well-defined species than the varieties of civilized countries.

Acknowledging the important role played by human selection, it becomes obvious how domestic organisms display adaptations of structure or habit conforming to human want or fancy. It also explains the frequently abnormal traits of domestic varieties and why their external characteristics vary greatly but their internal organs only slightly. Only with great difficulty can man select for structural deviations that cannot be externally observed, and he usually does not care for internal variations anyway. He can select only variations first provided by nature. He could not make a fantail, or think to or even try, until he found a pigeon with an unusually developed tail, or a pouter until he found a pigeon with an unusually large crop. The more abnormal a characteristic upon its first appearance, the more likely it is to catch his attention. But the phrase "to make a fantail" is not correct. The person who first selected a slightly larger tailed pigeon never imagined what its descendants would become through long-term, partly unconscious and partly methodical selection. Maybe the parent of all fantails had only fourteen slightly expanded tail feathers, like the Java fantail, or seventeen tail feathers like individuals of other breeds. Maybe the first pouter did not inflate its crop much more than the modern turbit inflates its upper esophagus, a habit disregarded by breeders because it is not one of the points of the breed.

A major structural deviation is not necessary to catch the breeder's eye, which perceives extremely small differences. It is human nature to value even a slight novelty in one's possession. The former value of slight differences cannot be judged by the value that such slight differences might have today after several breeds have been well established. Many slight deviations still arise among pigeons, but they are rejected as faults in the breed's perfection. The common goose has not given rise to any marked varieties, so it has recently been exhibited at poultry shows as distinct from the Thoulouse goose, from which it differs only in the most fleeting characteristic: color.

I think these ideas further explain why we know nothing about the origin or history of domestic breeds. In fact, a breed, like a dialect of a language, cannot be said to have a definite origin. A man preserves and breeds an individual with some slight structural deviation, or takes more care than usual in matching his best animals, thereby improving them, and the improved offspring slowly spread into the immediate neighborhood; they are not yet separately named, and their history is disregarded because they are only slightly valued. With further improvement by the same slow and gradual process, they spread more widely and are recognized as distinct and valuable, deserving of a local name. (In semicivilized regions with limited free communication, a new sub-breed will spread slowly.) As soon as the new sub-breed's valuable characteristics are fully acknowledged, unconscious selection always enhances the breed. Unconscious selection is perhaps more influential at one period than another, subject to the breed's popularity, and in one region than another, according to the state of civilization of the inhabitants. The chances of any record describing such slow and immediately unobservable changes are infinitely small.

A high degree of variability is obviously favorable to the human power of selection, because it provides the raw materials for selection to work on; not that mere individual differences are insufficient to allow for the accumulation of extensive modification in almost any desired direction. Manifestly useful or pleasing variations appear only occasionally, but the odds can be increased by keeping a large number of individuals, an important tool for success. Marshall has remarked about this principle with respect to sheep in parts of Yorkshire that "as they generally belong to poor people, and are mostly *in small lots,* they never can be improved." Professional plant breeders are generally more successful than amateurs in getting new and valuable varieties, because they raise large stocks of the same plant. When only a few individuals are kept, they are all allowed to breed, effectively preventing selection. Keeping large numbers of individuals naturally requires the creation of conditions favorable to that species to ensure proper breeding. But the most important point is that the organism should be so useful to or valued by humans that very close attention will be paid to even minor deviations in the quality or structure of each individual; without this, nothing can be effected. I have seen it gravely remarked how fortunate we are that the strawberry began to vary exactly when gardeners began to attend to it closely. Surely the strawberry had always varied, but slight variations were neglected; as soon as individual plants with slightly larger, earlier, or better fruit were selected for propagation over several generations (aided by some crosses with distinct species), those many admirable strawberry varieties appeared that have been raised for the past thirty or forty years.

Preventing crosses is an important element of success with animals that have separate sexes – at least in a region already stocked with other varieties. Enclosure of land plays a part in this. Wandering peoples or the inhabitants of open plains rarely have more than one breed per species. It is a huge convenience to the breeder and favorable to the formation of new breeds that pigeons mate for life, so many varieties can mingle in one aviary but keep true. Additionally, pigeons can be quickly propagated in large numbers and inferior birds easily rejected because they can serve as food when killed. Conversely, cats cannot be matched because of their nocturnal rambling habits, and though much loved by women and children, distinct breeds can hardly ever be maintained. The breeds that occasionally are observed are usually imported, often from islands. Although I do not doubt that some domestic animals intrinsically vary less than others, the rarity or absence of cat, donkey, peacock, goose, and other breeds is due mainly to selection not having been employed: in cats because they are difficult to pair, in donkeys because only a few are kept by the poor and little attention is paid to their breeding, in peacocks because they are difficult to rear and keep in large numbers, and in geese because they are valuable only for food and feathers, and because no one has taken pleasure in the display of distinct breeds.

To sum up the origin of domestic plant and animal varieties, environmental influence on the reproductive system is the most important cause of variability. I do not believe, as some authors do, that variability is an inherent and necessary contingency under all circumstances with all organic beings. The effects of variability are modified by inheritance and reversion. Variability is governed by many unknown rules, especially correlated growth. The direct action of the environment, and use and disuse, may have some effect. The final result is thus infinitely complex. In some cases intercrossing of originally distinct species played an important part in the origin of domestic breeds. The occasional crossing of established domestic breeds contributes to the creation of sub-breeds with the aid of selection, but the importance of crossing varieties has been greatly exaggerated both in animals and plants propagated by seed. (The importance of crossing distinct species and varieties is immensely important for plants temporarily propagated by cuttings, buds, etc., because the cultivator disregards the extreme variability of hybrids and mongrels and the sterility of hybrids. Plants not propagated by seed are temporary, so they are of little importance to this discussion.)[7] Over all these causes of change, the cumulative action of selection, whether applied methodically and relatively quickly or unconsciously and relatively slowly but more efficiently, is by far the most predominant power.

7. [Grafting and certain other forms of plant propagation do not involve reproductive cells or any other means for the exchange of (what we now know to be) genetic material, so these methods do not actually involve "crossing." – D.D.]

2

VARIATION IN NATURE

BEFORE APPLYING THE PRINCIPLES FROM THE LAST CHAPTER to living things in the wild, we need to establish whether or not they too are subject to variation. A proper treatment of this topic would involve a long catalog of dry facts, but I will reserve this for my future work. And I won't discuss the various definitions of "species," because no one definition satisfies all naturalists, even though everyone vaguely knows what it means. Generally, the term includes an unknown element of a distinct act of creation. "Variety" is almost as difficult to define, but in this case community of descent is often implied even though it can rarely be proven. There are also monstrosities – by which I mean considerable structural deviations that are either harmful or useless to the species and not usually propagated – but these graduate into variations. Some authors use the term "variation" in a technical sense to indicate an un-inheritable modification resulting from environmental conditions, but the dwarfed shells of brackish Baltic waters, dwarfed plants of alpine summits, and the thickened fur of animals living far north might be inherited for at least a few generations, and in such cases I would call the form a variety.

Once again, consider that there are many slight individual differences, such as those observed in offspring from the same parent. (Sometimes differences are just assumed to have arisen this way because they are frequently observed in individuals of the same species living in a confined locality.) No one would argue that all the individuals of a species are cast in the very same mold. Individual differences are important because they supply material for natural selection to accumulate, just as humans can accumulate individual differences in any direction in domestic organisms. Individual differences generally occur in what naturalists consider "unimportant parts," but I could show a long list of examples where physiologically or classificatorily important parts also vary. I am certain that the most experienced naturalist would be surprised at how many cases of variability, even in important structural parts, can be collected from good authorities. Keep in mind that systematists do not like finding variations of important characteristics, and there are not many people prepared to laboriously compare the internal organs of many specimens from one species. I never expected that the branching pattern of nerves close to an insect's central ganglion would vary within a species – I expected such changes to occur in small increments – yet Mr. Lubbock recently showed the variability of these main nerves in *Coccus* to be comparable to the irregular branching of trees. He also found that the muscles in the larvae of certain insects are far from uniform. Some authors state that important organs never vary, and in a circular argument they basically classify "important traits" as those that are invariable (a few naturalists have honestly confessed this). In this scheme there cannot be any varying important parts by definition! But under any other scheme many examples can be given.

With respect to individual differences, one point strikes me as extremely perplexing: the "polymorphic" genera, in which species vary so inordinately that few naturalists can agree which forms should be ranked as species and which as varieties. Among plants, examples include *Rubus, Rosa,* and *Hieracium;* among animals, there are several genera of insects and brachiopods. Most polymorphic genera contain some species with fixed characteristics. Also, genera that are polymorphic in one region seem to be polymorphic in others, with a few exceptions. Judging from brachiopod shells, genera remain polymorphic through time. These observations are perplexing because they suggest that this type of variability is independent of the environment. I suspect that members of polymorphic genera vary in characteristics that are of no service or disservice and have therefore not been seized and rendered definite by natural selection, as I explain later.

Those forms that could justifiably be considered species but are similar enough to other forms or linked to them by intermediates so that naturalists do not rank them as distinct species are very important for this discussion. There is every reason to believe that many of these "doubtful" forms have permanently retained their characteristics in their native regions for as long as "true" species. On a practical level, when a naturalist can link two forms through intermediates, he will treat the most common one, or the one that happened to have been described first, as the species and the other as a variety. But there are some very challenging cases – I won't list them – concerning whether or not one form should be classified as a variety of another even if they are closely linked by intermediates. The commonly assumed "hybrid nature" of intermediates does not always solve the problem. In many cases the intermediate forms have not been found, and they are assumed by analogy to either exist undiscovered or to have become extinct; here a wide door for the entry of doubt and conjecture is opened.

It cannot be disputed that these doubtful forms are common.[1] A surprising number of plants from Great Britain, France, and the United States have been ranked as species by some botanists and mere varieties by others. Mr. H. C. Watson, to whom I am grateful for all kinds of assistance, has listed for me 182 British plants that are generally considered varieties but have all been ranked by some botanists as species. He did not include many minor varieties that have nevertheless been ranked as species by some botanists; he entirely omitted several highly polymorphic genera. Under genera, Mr. Babington lists 251 species, whereas Mr. Bentham lists only 112, amounting to a difference of 139 doubtful forms! Doubtful forms of mobile animals that mate to reproduce are rare within the same region but common in separated areas. Many slightly differing North American and European birds and insects are ranked by one naturalist as undoubted species and by another as varieties. Many years ago, when comparing birds of the Galápagos Archipelago to one another and to those of the American mainland, I was struck by the

---

1. The sound opinion and wide experience of naturalists seems the only guide to follow in determining whether a form should be ranked as a species or a variety. However, in many cases the best course is to follow the majority, because there are very few well-defined and known varieties that have not also been listed as species by at least some qualified judges.

vague and arbitrary distinction between species and varieties. Many of the insects on Madeira are ranked as varieties in Mr. Wollaston's admirable book but would certainly be characterized as species by many entomologists. Even Ireland has a few animals generally considered varieties but ranked as species by some zoologists. Several experienced ornithologists consider the British red grouse a variety of a Norwegian species, while most rank it as a species peculiar to Great Britain. If there is a large distance between the homes of two doubtful forms, many naturalists rank them as species, but what distance will suffice? If that between Europe and America is enough, is the distance between the Continent and the Azores, or Madeira, or the Canaries, or Ireland also enough? Trying to discuss the demarcation between a variety and a species before establishing any definition of these terms is like hammering the air.

Several interesting arguments from geographic distribution, analogical variation, hybridism, and so forth have been developed to address this problem with respect to different cases. I will discuss only the example of the primrose and cowslip. These plants differ in appearance, flavor, odor, flowering period, and range. They grow in somewhat different habitats and ascend mountains to different heights. Finally, according to many experiments conducted over the course of several years by the careful observer Gärtner, they can be crossed only with great difficulty. There could hardly be better evidence that two forms are distinct species. But for all that, they are linked by many intermediates that are probably not hybrids. There is also overwhelming experimental evidence showing that they have descended from a common parent and are therefore varieties.

In most cases, close investigation brings naturalists to agree on how to rank a doubtful form. Yet the greatest number of doubtful forms is found in the best-known regions. I have noticed that if any plant or animal is useful or interesting to humans, varieties of it will be found recorded; these varieties, moreover, will be ranked by some authors as species. Consider the common oak, so closely studied. A German author makes more than a dozen species out of forms generally considered varieties. And in this country some of the highest botanical authorities claim sessile and pedunculated oaks as species while others claim them as varieties.

When a young naturalist begins studying an unfamiliar group of organisms, he will find it difficult to determine differences that delineate species and those that delineate varieties, because he doesn't yet know the amount and kind of variation within the group. (This demonstrates, at least, that there is often some variation.) But if he confines his attention to one class in one region, he will soon make up his mind about how to classify doubtful forms. He will define many species because the differences impress him, and he lacks a general knowledge of analogical variation in other groups and other countries. As he extends his range of observation, he will discover more cases that are difficult to classify, for he will encounter more forms that are closely related. If his observations are extended further, he will ultimately make up his mind about species and varieties, but only at the expense of admitting extensive variation – an admission other naturalists will dispute. When he comes to study related forms from regions that are no longer continuous, and where intermediate links between doubtful forms will necessarily be absent, he will have to trust almost entirely to analogy and his difficulties will rise to a climax.

No clear line has been drawn between species and sub-species (forms that are nearly ranked as species but not quite there), sub-species and varieties, or lesser varieties and individual differences. They all blend into each other in a smooth series, and a series impresses the mind with the idea of an actual progression.

Although individual differences are unimportant to the systematists, they are very important to this argument as the first step toward the minor varieties barely thought worthy of notice in annals of natural history. I consider even slightly distinct and permanent varieties as steps leading to still more distinct and more permanent varieties, and these in turn as leading to sub-species and species. Progression from one stage to another may in some cases be due merely to the continuous and long-term action of different environmental conditions in two different regions, but I have little faith in this view. Instead, when a variety changes from differing little from its parent to one that differs more, I attribute the change to natural selection having accumulated structural differences in certain definite directions. *Therefore, I believe a distinct variety can be called an incipient species.* Whether this is justified should be judged by the general weight of observations and ideas presented throughout this book.

Not all incipient species actually become species. They may become extinct in the incipient state or simply remain varieties for very long periods (as Mr. Wollaston has shown for certain mollusk varieties based on fossil land shells in Madeira). If a variety were to flourish and exceed its parent species in number, it would be ranked as a species and the parent as a variety; it might supplant and exterminate the parent or coexist with it, both being ranked as independent species. (I will return to this subject.)

From these considerations I take the term "species" as being arbitrarily assigned for the sake of convenience to groups of individuals closely resembling one another. It does not essentially differ from the term "variety," given to less distinct and more fluctuating forms. "Variety" is also a term that is applied arbitrarily and for convenience, again, in comparison with mere individual differences.

Guided by theoretical considerations, I thought that tabulating all the varieties of several well-studied plants would yield some interesting data about the nature of species that vary the most. At first this seemed simple, but Mr. H. C. Watson and Dr. Hooker soon convinced me of the many difficulties. (I am reserving the tables themselves and a discussion of these difficulties for my later work.) After carefully reading my manuscript and examining the tables, Dr. Hooker agreed that the following statements are fairly well established. Nevertheless, the subject is complex and treated tersely here, so allusions cannot be avoided to "the struggle for existence," "divergence of character," and other concepts that will be discussed later.

Alph. de Candolle and others have shown that plants with extensive ranges generally have varieties. This might have been expected because it means exposure to diverse physical conditions and competition with different sets of organisms (competition is by far the more important, as discussed later). My tables also show that in a limited region the most common species (those with the greatest number of individuals) and the species that are most diffused often have varieties well-defined enough to be recorded in botanical works.[2] Therefore the most "dominant" species – those that range most extensively over the world, are the most diffused in their native regions, and are the most numerous – are the ones that most often produce varieties, or as I consider them, incipient species. This too might have been anticipated, because for a variety to become permanent, it must struggle with other inhabitants of the region. Already dominant species are the most likely to leave offspring, which, in addition to being slightly modified, will inherit those advantages that enabled their parents to become dominant.

If the plants of a particular region are separated into two groups, with those belonging to large genera in one and those belonging to small genera in the other, then a greater number of dominant species will be found on the side of larger genera. Again, this might have been anticipated: if many species of a genus inhabit a region, it shows that something about the organic or inorganic conditions of that region are favorable to that genus; therefore a large genus encompassing many species contains a proportionally greater number of dominant species. But so many factors tend to obscure this result that I am surprised my tables show even a small majority of dominant species among large genera. To give examples of how complex this is, consider that freshwater and salt-loving plants are generally diffused with extensive ranges, but this is probably because of the nature of their habitats rather than the size of the genera to which they belong; likewise, the wide distribution of simple plants is also unrelated to genera size. (The reason that simple plants range extensively will be discussed in the chapter on geographic distribution.)

Because I think of species as being just well-defined varieties, I anticipated that species from a large genus would tend to have more varieties than species from a small genus: if many closely related species (i.e., members of a genus) have already formed, many incipient species should still be forming. Where many large trees grow, we expect to find saplings. Where many species within a genus have arisen through variation, circumstances have favored variation, and we expect them to still

2. Being "diffused" is a different consideration from "extensive range" and "commonness."

favor variation. However, if we consider each species a special act of creation, then there is no apparent reason to expect more varieties in a large genus than in a small one.

To test this I arranged the plants of twelve countries and the coleopterous insects of two districts into two groups with, again, species from larger genera on one side and those from smaller genera on the other. The species in the group of large genera invariably have a greater proportion of varieties than species in the group of small genera. Moreover, species of large genera with any varieties invariably have a higher average number of varieties than species of small genera. The same results follow when the division is made after excluding genera with four or fewer species. These observations are understandable if species are just well-defined and permanent varieties, because wherever many species within a genus have been generated, the generation should still be happening, especially because the process of species formation is probably slow. This is in fact the case if varieties are considered incipient species. This is not to say that all large genera vary greatly, their ranks now swelling with species; moreover, some small genera do vary and are increasing. It would be fatal to my theory if this were not so, because geology plainly reveals that small genera have often increased greatly and large genera have peaked, declined, and disappeared.[3] All I want to show is that if many species have been formed within a genus, on average, many are still forming – and this is supported by my data.

Again, there is no infallible criterion by which a species can be distinguished from a well-defined variety, and if intermediate links between doubtful forms cannot be found, then naturalists are compelled to come to a determination based on the amount of difference between them and judge by analogy whether it suffices to rank one, the other, or both as species. Fries has remarked with respect to plants and Westwood with respect to insects that in large genera the differences between species are exceedingly small. I have tried to test this numerically by averages, and as far as my imperfect results go, they always confirm it. I also consulted several knowledgeable and experienced observers who, after delibera-

3. [Note that in Darwin's time, geology encompassed what we would today recognize as paleontology. – D.D.]

tion, concurred. Therefore, species in large genera resemble varieties more than do the species of small genera. Put another way, large genera with many incipient species contain already-generated species that resemble varieties because they do not differ from one another extensively.

Moreover, species within a large genus are related to one another in the same way as varieties within a species. Of course, species within a genus are not all equally distinct from one another; they can generally be divided into sub-genera, sections, or lesser groups. Fries has correctly remarked that little groups of species cluster like satellites around certain other species, and what are varieties but unequally related groups of forms clustered around their parent species? There is one important difference between varieties and species: the amount of difference between varieties when compared to one another or their parent species is less than the amount of difference between species within a genus. This will be explained in the discussion of what I call "divergence of character," along with the tendency of varietal differences to burgeon into the greater differences between species.

I think one other point is worth noting. Varieties generally have restricted ranges (although this is a truism, for if a variety were found to have a greater range than its parent species, their denominations would be reversed). But species that are related to many other species, and therefore resemble varieties, also tend to have restricted ranges. For example, Mr. H. C. Watson has identified for me sixty-three species-ranked plants in the fourth edition of the well-sifted *London Catalogue of Plants* that he considers so closely related to other species as to doubt their assigned rank. These sixty-three supposed species range on average over 6.9 of the provinces into which Mr. Watson has divided Great Britain. The same catalog lists fifty-three acknowledged varieties that range over 7.7 provinces, whereas the species to which they belong range over 14.3 provinces. This means that the acknowledged varieties have almost the same average range as the forms Mr. Watson identifies as doubtful species but that British botanists almost universally mark as true species.

Finally, then, varieties and species have the same general properties and cannot be distinguished except by (1) the discovery of intermediate links, although such links do not affect the characteristics of the forms they connect, and (2) a certain amount of difference that cannot be exactly defined. Within any given region, genera with a greater than average number of species also have species with a greater than average number of varieties. In large genera, species tend to be closely but unequally related and clustered around certain species. Species very closely related to other species apparently have restricted ranges. In all of these respects, species of large genera are like varieties. These patterns can be clearly understood if species were once varieties, but they are completely inexplicable if each species has been independently created.

Also, on average, flourishing and dominant species of large genera vary the most, and varieties, as discussed later, tend to become converted into new and distinct species. Large genera thus tend to become larger and dominant forms become more dominant as they leave many modified and dominant descendants. Yet by steps explained later, large genera also tend to break up into smaller genera. And so it is that the life forms of the universe become divided into groups subordinate to groups.

3

THE STRUGGLE FOR EXISTENCE

BEFORE DISCUSSING THE STRUGGLE FOR EXISTENCE, I NEED to show how it is relevant to natural selection. As mentioned in the previous chapter, individual organisms in the wild vary from one another. (I am not aware that this has ever been disputed.) It is not important whether a multitude of doubtful forms are called "species," "sub-species," or "varieties." For example, what rank the two or three hundred doubtful British plants are entitled to hold is immaterial if the existence of any well-marked varieties is accepted. The existence of individual variability and of well-marked varieties is a necessary foundation, but does not help explain *how* species arise. How have all the exquisite adaptations of one part of the organization to another and of living things to their environments – including other living things – been perfected? We see beautiful coadaptations in the woodpecker and mistletoe, in the humblest parasite that clings to the hairs of a quadruped or the feathers of a bird, in the structure of a beetle that dives through water, and in the plumed seed that is wafted by the gentlest breeze. In short, we see beautiful adaptations in every part of the organic world.

How are varieties – incipient species – converted into distinct species? How do groups of species making up distinct genera arise? All of these results follow from the struggle for life. Any variation, regardless of magnitude or ultimate cause, that is profitable to an individual through its complex relationships with other organisms and nature increases that individual's chances of survival and will generally be inherited by its offspring. These offspring will consequently also have a better chance of surviving, because only a small number of individuals that are born can

actually survive. I call this principle, by which each slight useful variation is preserved, "natural selection" to mark its relation to the human
power of selection. I have shown that by selection, humans can produce
great results and adapt organisms to their own uses by accumulating
slight but useful variations provided by nature. But as I will show, natural selection is a power that is always in action and is as immeasurably
superior to feeble human efforts as the works of nature are to those of art.

Sir Charles Lyell and the elder de Candolle have shown that all organisms face severe competition. With respect to plants, W. Herbert,
Dean of Manchester, has treated this subject outstandingly, evidently
the result of his great horticultural knowledge. Nothing is easier than
recognizing the universal struggle for life, but nothing is more difficult
than constantly keeping it in mind. Yet unless it becomes thoroughly in-
grained, I am convinced that the whole economy of nature – distribution,
rarity, abundance, extinction, and variation – will be seen only dimly or
misunderstood. We behold the face of Nature bright with gladness and
often see a superabundance of food, but we forget that the birds singing
idly around us live on insects or seeds, thus constantly destroying life;
we forget how frequently these songsters, their eggs, and their nestlings
are destroyed by birds or other predators; we forget that although food
may be plentiful now, it is not so in all seasons or in all years.

I use the term "struggle for existence" broadly and metaphorically to
include the dependence of organisms on one another, and, more importantly, the success an individual has in leaving progeny. In a time of want,
two dogs may truly be said to struggle with each other over which shall
get food and live. But a plant on the edge of a desert struggles against
drought; it is dependent on moisture. A plant that produces a thousand
seeds a year, of which on average only about one will mature, struggles
against members of its own species and other plants that already clothe
the ground. The mistletoe is dependent on the apple and a few other
trees but does not exactly struggle with them – yet if too many of these
parasites were to cover one tree, it would languish and perish. However,
several seedling mistletoes growing together on one branch do struggle
with one another. The mistletoe struggles metaphorically against other
fruit-bearing plants in tempting birds that devour and disseminate seeds.
I use the term "struggle for existence" to cover all these cases.

A struggle for existence inevitably follows from the high rate at which all living things tend to multiply. Every reproducing organism must suffer destruction at some point; otherwise its numbers would swell geometrically to such inordinate proportions that no region could support them. Because more individuals are produced than can possibly survive, there must be a struggle for existence – either one individual with another of the same species, or with individuals of other species, or with the physical environment. This is the doctrine of Malthus applied to the whole organic world; for here there can be no artificial abundances of food or prudent restraint from mating. Although some species may now be increasing in numbers, they cannot all increase, because the world would not hold them.

Every species, without exception, increases at such a high rate that without checks the earth would become covered by the progeny of a single pair. Even slow-breeding humans have doubled in twenty-five years. At this rate, in a few thousand years there would literally be no standing room. Linnaeus calculated that if an annual plant produced only two seeds – and no plant is as unproductive as this – and their seedlings produced two seeds each, and so on, then in twenty years there would be a million plants. The elephant is thought to be the slowest breeder of all known animals, and I have tried to estimate its minimum rate of natural increase. Conservatively estimating that elephants breed from age thirty through age ninety, producing three pairs of young during the interval, then after five hundred years there would be fifteen million *living* elephants descended from the first pair.

But there is better evidence than just theoretical calculations: the numerous cases of animals in the wild increasing at astonishing rates when conditions were favorable for several consecutive seasons. Even more striking are cases of domestic animals that have run wild. Had they not been confirmed, rumors of slow-breeding cattle and horses in South America and Australia increasing at great rates would have been incredible. Similarly, there are examples of invasive plants becoming common across whole islands in less than ten years. Several of the plants covering entire square leagues of the plains of La Plata, almost to the exclusion of all other plants, were introduced from Europe. I hear from Dr. Falconer that several American plants introduced to India now range from Cape Comorin to the Himalaya – a change that has happened in less than four hundred years. The fertility of these plants and animals had not been suddenly and temporarily increased. The obvious explanation is that the environments were very favorable, so fewer old and young were destroyed and many of the young could reproduce. The geometric ratio of increase, which always produces surprising results, explains the extraordinarily rapid growth and spread of invasive organisms.

In the wild almost every plant produces seed and almost every animal pairs annually, which means that the potential for geometric growth is widespread – but held in check. Familiarity with large domestic animals tends, I think, to mislead us. We do not see them succumb to great destruction, and we forget that thousands are slaughtered each year for food. In the wild an equal number would somehow have to be destroyed.

The only difference between organisms that produce eggs or seed by the thousand and those that produce extremely few is that the less productive would require a few more years of favorable conditions to populate an entire region. The condor lays a couple of eggs and yet may be more numerous than the ostrich, which lays twenty. The Fulmar petrel lays only one egg but is believed to be the most numerous bird in the world. One fly deposits hundreds of eggs and another (like the louse fly) deposits only one, but this does not determine how many individuals of each are supported in a region. Producing many eggs is important to species that are dependent on fluctuating quantities of food, because it allows for quick increases in number; however, the real importance of laying many eggs is to make up for the destruction that happens at some, usually early, period of life. If an animal can protect its eggs or young, then average numbers can be maintained even if only a small number of offspring are produced. But if many eggs or young are destroyed, many have to be produced, otherwise the species would become extinct. If a tree lived, on average, for a thousand years, then a single seed every thousand years would sufficiently maintain its numbers – assuming that every single seed were to survive and germinate in an appropriate place. So the average number of any animal or plant depends only indirectly on the number of its eggs or seeds.

These considerations are essential when thinking about nature: every organism strives to increase its numbers, each individual struggles at some period of its life, and destruction inevitably falls on the young or the old during each generation or at recurrent intervals. Lighten any check, mitigate the destruction ever so little, and the number of individuals will quickly increase. The face of Nature is like a yielding surface with ten thousand sharp wedges packed close together and driven inward by incessant blows; sometimes one wedge is struck, and sometimes another with even greater force.

What moderates the natural tendency to increase is obscure. Consider the most vigorous species; by as much as it swarms in numbers, so much greater will be its tendency to increase. There isn't a single case for which we know exactly what the checks are. This should not surprise anyone who recognizes how little we know about this topic even with respect to humans, so much better known than any other animal. Several authors have treated this subject, and in my future work I will discuss some of the checks at length, especially those influencing the feral animals of South America. Here I mention only the main points. Eggs or very young animals seem to suffer the most, but this is not always the case. Plants face a vast destruction of seeds, but based on some observations I have made, the seedlings suffer the most from germinating in ground already thickly covered by other plants. Seedlings are also destroyed in vast numbers by various enemies. I dug and cleared a three-by-two-foot piece of ground where there could be no choking by other plants and marked all the seedlings of native weeds as they came up. Out of 357 seedlings, 295 were destroyed, mostly by slugs and insects. If a piece of turf is mowed or browsed by ruminants over a long period of time and then allowed to grow uninhibited, the more vigorous plants gradually kill the less vigorous (although fully mature) plants. In this way, on a little three-by-four-foot plot of turf, nine species out of twenty perished because all were allowed to grow freely.

Of course, the amount of available food defines the upper limit to a species' rate of increase, but frequently it is predation that determines its average number. The stocks of partridge, grouse, and hare on large estates depend chiefly on the destruction of vermin. If in England not a single game animal were shot for the next twenty years, and if at the same time no vermin were destroyed, there would probably be *less* game than now (even though hundreds of thousands of game animals are killed annually). However, in some cases there is no destruction by predators – as with the elephant and rhinoceros. Even the tiger in India rarely dares to attack a young elephant protected by its dam.

Climate plays an important part in determining the average number of a species; periodic extremes of cold or drought are the most effective checks of all. I estimate that the winter of 1854–1855 destroyed four-fifths of the birds on my own grounds. This is tremendous – a human epidemic that kills 10 percent is considered extraordinarily severe. At first sight the action of climate seems independent of the struggle for existence, but insofar as climate reduces food supplies, it can precipitate the most intense competition among individuals – be they of the same or different species – that subsist on the same kind of food. Even when climate acts directly – for example, through extreme cold – it is the least vigorous or those with the least food that suffer most. When we travel from south to north or from a humid region to a dry region, we invariably see some species gradually disappear. It may be tempting to attribute the whole effect to the conspicuously changing climate and its direct action, but this would be incorrect. Even where a species is most plentiful, it constantly suffers enormous destruction at some period of its life from predators or from competitors for space and food. If these enemies or competitors are favored by any slight changes in climate, they will increase. And because each area is already fully stocked by inhabitants, the other species will decrease. So when we travel southward and observe fewer and fewer individuals of some species, the cause involves both the species in question being hurt and others being favored. When we travel north, the effect is similar but less pronounced, because all kinds of species, and therefore competitors as well, decrease northward. When going north or ascending a mountain, it is more common to meet with stunted forms due to the directly detrimental action of climate. In Arctic regions, snow-capped mountains, or absolute deserts, the struggle for life is almost exclusively with the elements.

Thus, the action of climate is mainly indirect by favoring other species. For example, there are prodigious numbers of garden plants that can perfectly well endure our climate but never become naturalized, because they cannot compete with native plants or evade destruction by native animals.

Epidemics tend to ensue when a species increases inordinately within a small area due to very favorable circumstances. (At least this seems to be generally true for game animals.) This is a limiting check independent of the struggle for life. But even some of these so-called epidemics appear to be caused by parasitic worms that have been disproportionately favored for some reason (possibly because they spread more easily among crowded animals); this produces a sort of struggle between the parasite and its host.

In many cases preservation of a species depends on the maintenance of a large number of individuals relative to its enemies. It is easy to raise corn, rapeseed, and other grains in a field, because the seeds far outnumber the birds that feed on them. The birds cannot increase proportionally to this superabundance of food in one season, because their numbers are checked by winter. But anyone who has tried knows the difficulty of getting seeds from a few wheat or other such plants in a garden; in my case I lost every single seed. The necessity for a large stock in some species explains some singular natural phenomena, such as the extreme abundance of otherwise rare plants in certain areas and the density of some "social" plants even at the extremes of their range. In such cases a plant will survive only where conditions are so favorable that many individuals can exist together and save one another from destruction. I'll add, without elaboration for now, that the positive effects of frequent intercrossing and the negative effects of close inbreeding are probably relevant to some of these examples.

Many recorded cases show the unexpected and complex relationships among organisms that have to struggle together in one region. I will give a simple but interesting example. On the estate of a relative in Staffordshire, there is a large, barren, and untouched heath. Several hundred acres of it were enclosed and planted with Scotch fir twenty-five years earlier. In the planted part, the native vegetation changed remarkably, more so than is generally observed when passing from one soil to another. Not only did the proportional numbers of heath plants change, but twelve species (not counting grasses and carices) that were absent from the heath flourished on the plantation. The effect on insects must have been even greater, because the heath was frequented by two or three insectivorous bird species, but the plantation harbored six bird species not found on the heath. Here we see how introducing a single type of tree had potent effects; the only other interference was enclosure of the land to keep out cattle. Indeed, I observed the importance of enclosure near Farnham, in Surrey, where there are extensive heaths with a few clumps of Scotch fir on distant hilltops. In the last ten years large spaces have been enclosed, and self-sown firs are now springing up in multitudes, so densely that all cannot survive. When I ascertained that these young trees had not been sown or planted, I was surprised by their numbers; I went to several places where I could see hundreds of acres of unenclosed heath and saw literally no Scotch firs except for the old planted clumps. But on looking closely I found many seedlings and little trees that were perpetually browsed down by cattle. In one square yard, several hundred yards from one of the old clumps, I counted thirty-two little trees. Judging from growth rings, one of them had tried to raise its head above the shrubs of the heath for twenty-six years and failed. No wonder that as soon as the land was enclosed, it became thickly covered with firs. Yet the heath was so barren and extensive that no one would have suspected cattle of having searched it for food so effectively!

In this case cattle determined the existence of Scotch fir, but in some parts of the world insects determine the existence of cattle. Paraguay offers perhaps the most curious example. Here cattle, horses, and dogs have never run wild, and yet to the south and to the north they swarm in a feral state. Azara and Rengger have shown that this is caused by an abundance in Paraguay of a certain fly that lays its eggs in the navels of these animals when they are born. These flies are numerous and their increase must be checked, probably by birds. So if certain insectivorous birds – whose numbers are probably regulated by hawks and other predators – were to increase in Paraguay, the flies would dwindle, the cattle and horses would become feral, the vegetation would change (as I have observed in some parts of South America), the insects and thus the insectivorous birds (as in Staffordshire) would be affected, and so on, in ever greater circles of complexity. This series begins and ends with insectivorous birds – not that natural relationships are ever so simple. Battle within battle must be perpetually recurring, with varying success, yet in the long run, forces are balanced and the face of Nature remains uniform over long periods of time, although the slightest change would give victory to one organism over another. Our ignorance is so profound and our presumptions are so high that we marvel at the extinction of a species; failing to see the cause, we invoke cataclysms to desolate the world and invent laws to govern the duration of forms of life!

I will give one more example showing how plants and animals widely separated on the scale of nature are bound together by a web of complex relationships.[1] The exotic and peculiarly structured *Lobelia fulgens* is never visited by insects in this part of England and therefore never sets seed. (Many orchids absolutely require moths for pollination. Comparably, bumblebees are indispensable for the pollination of heartsease, because other bees do not visit this flower.) My experiments show that if bees are not essential to the pollination of clovers, they at least help significantly. But the common red clover is visited only by bumblebees, because no other bees can reach the nectar. So I have little doubt that if the bumblebee became very rare or extinct in England, the heartsease and red clover would follow. The number of bumblebees in a region depends greatly on the number of field mice, which destroy their combs and nests. Mr. H. Newman has long studied bumblebees and believes that "more than two-thirds of them are thus destroyed all over England." Now, everyone knows that the number of mice depends mostly on the number of cats: Mr. Newman says, "Near villages and small towns I have found the nests of bumblebees more numerous than elsewhere, which I attribute to the number of cats that destroy the mice." Therefore, the presence of many cats in a region may determine – through mice and then through bees – the frequency of certain flowers!

Every species faces multiple checks acting at different periods of life and different seasons or years. One or a few of these are generally the most potent, but all conspire in determining the average number or even existence of a species. In some cases widely disparate checks act on the same species in different regions. Looking at the plants and bushes on an entangled bank, we are tempted to attribute the kinds of plants and their proportional numbers to "chance," but that would be wrong! When an American forest is cut down, a very different vegetation springs up, but the trees now growing on ancient Indian mounds in the southern United

---

1. [The "scale of nature" is an archaic system of ordering all beings, single-file, from the "highest" to the "lowest." Darwin and his naturalist contemporaries seem to have used it in a less strict sense to characterize organisms based on complexity. – D.D.]

States are as diverse as in the surrounding virgin forests. A momentous struggle must have ensued for long centuries between various types of trees, each annually scattering seeds by the thousands. What war must have gone on between insect and insect, between insects, snails, and other animals with birds and beasts of prey, all striving to increase and all feeding on one another, trees, seeds, seedlings, or the other plants that first covered the ground and checked the trees' growth! Throw up a handful of feathers and they all fall to the ground according to definite laws, and yet that is a very simple problem when compared to the interactions that determined, over centuries, the kinds and proportional numbers of trees now growing on the old Indian ruins!

Organisms that depend on each other, as a parasite depends on its prey, are generally far apart on the scale of nature. This is particularly common with those directly engaged with each other in a struggle for existence, such as locusts and grass-feeding ruminants. But the struggle is usually the most severe between members of the same species, because they inhabit the same regions, require the same food, and are exposed to the same dangers. The struggle is usually just as severe between varieties of the same species, and sometimes the contest ends quickly. If several varieties of wheat are sown together and the resulting seed is mixed and resown, the most fertile varieties or those best suited to the soil or climate will grow better, yield more seed, and in a few years supplant the others. To keep up a mixed stock of different-colored sweet peas, which are very close varieties, requires separate harvesting and manual mixing of seeds to the desired proportions; otherwise the weaker kinds will steadily decrease in number and disappear. Similarly, it is said that certain varieties of mountain sheep cannot be kept together because one will starve out the others. The same result follows when keeping different varieties of medicinal leeches together. It is unlikely that varieties of any domestic plants or animals have so exactly the same strengths, habits, and constitutions that the original proportions of a mixed stock could be kept up for half a dozen generations if they were allowed to struggle as in the wild and without annual sorting of the seed or young.

Because species within the same genus are always similar in structure and usually in habit and constitution, the struggle between its members will generally be more severe than between members of different genera when they come into competition with one another. The recent extension of one swallow species in the United States caused the decrease of another swallow species; the recent increase of mistle thrush in parts of Scotland caused the decrease of the song thrush; across Russia, the small Asiatic cockroach has driven out a larger species; one species of charlock will supplant another; and we often hear that in many different climates one rat species takes the place of another. We can dimly see why competition should be most intense between closely related forms that fill similar niches. But there is probably no case in which we can know precisely why one species has been victorious over another in the great battle of life.

An important corollary can be deduced from these remarks: the structure of every organism is related in an essential but often hidden way to that of every other organism with which it competes for food or space, it preys on, or it must escape from. This is obvious in the structures of the teeth and claws of the tiger, and in the legs and claws of the parasite clinging to its hair. In the beautifully plumed seeds of the dandelion and in the flattened and fringed legs of the water beetle, the relationship seems at first confined to the elements of air and water. But the advantage of plumed seeds is no doubt due to the land already being thickly crowded by plants, so such seeds can be broadly distributed and fall on unoccupied ground. A water beetle can compete with other aquatic insects, hunt prey, and elude predators because the structure of its legs is so well adapted for diving. The store of sustenance in seeds may appear to have no relationship to other plants, but judging by the strong growth of young plants produced from such seeds (like peas and beans), I suspect that large stores of sustenance favor the growth of young seedlings struggling with vigorous plants already growing all around.

Why doesn't a plant in the middle of its range double or quadruple in number? It can clearly withstand slightly warmer, colder, dryer, or damper conditions, because in other places it ranges into such areas. If we want the plant to proliferate in this hypothetical case, we must endow it with some advantage over its competitors or the animals that feed on it. A constitutional change that helps it endure climate may suffice at the periphery, but only a few plants or animals range so far that that they are destroyed by climate alone. Only in extreme conditions, like in the Arctic or a total desert, will competition cease. Even in extremely cold or dry lands there is competition between a few species or between individuals of the same species for the warmest or dampest spots.

Thus if a plant or animal is transplanted to a new region with new competitors, its environment will be fundamentally different even if the climate is identical to that of its former home. The hypothetical modification necessary to increase its average numbers is not the same in the new location as in the old; some advantage over different competitors or enemies is needed.

It is useful to try to imagine how one form could be endowed with an advantage over another, even though there is probably no case where we would know what to do so as to succeed. This exercise will convince us of our ignorance of the mutual relationships among organisms, a conviction as necessary as it is difficult to acquire. Keep in mind that each organism strives to increase geometrically; that each at some period of its life, during some season of the year, or during each generation or at intervals, struggles for existence, and suffers great destruction. When we reflect on this struggle, we may console ourselves with the belief that the war of nature is not incessant, that no fear is felt, that death is generally prompt, and that the vigorous, healthy, and happy survive and multiply.

4

NATURAL SELECTION

HOW DOES THE STRUGGLE FOR EXISTENCE INFLUENCE VARIA-tion? Does selection – so potent in human hands – apply in nature? I think it does, most effectively. Recall the strength of heredity and the endless peculiarities in domesticated organisms, and to a lesser extent in wild organisms. (Under domestication the whole organization becomes somewhat plastic.) Also recall the complex and close-fitting relationships of all organisms to one another and to their physical environments. If variations useful to humans have occurred, then surely variations useful to each organism in the great and complex battle of life also sometimes occur in the course of thousands of generations. Accepting this and adding that many more individuals are born than can possibly survive, can it be doubted that those with even a slight advantage will have the best chance of surviving and propagating their kind? Moreover, it is certain that even slightly detrimental variations are destroyed. I call this preservation of favorable variations and rejection of detrimental variations "natural selection." Variations that are neither useful nor detrimental are not affected by natural selection and remain a fluctuating element, perhaps as observed in so-called polymorphic species.

We can best understand the probable course of natural selection by considering a region undergoing some physical change, say, in climate. The relative proportions of different organisms would change almost immediately and some species might go extinct. As I discussed in the previous chapter, if the population of one species changes, many others will be affected, independently of climatic shifts, because a region's inhabitants are intricately linked. If the borders of this region are open, then new forms would immigrate, also disturbing the relationships among some of the original inhabitants. Remember how extensive the impact of a single introduced tree or animal can be. But in the case of an island or a region partly surrounded by barriers, into which new and better-adapted forms could not enter, there would be niches that would be better filled if some of the original inhabitants were modified. If instead the area had been open to immigrants, these niches would have been seized by the intruders. Then, in the course of ages, every slight modification that chances to arise and that gives some individuals an advantage over the others in better adapting them to the altered conditions will tend to be preserved. Natural selection therefore has free scope for the work of improvement.

As stated in the first chapter, there is reason to believe that changes in the environment cause or increase variability by acting on the reproductive system. In the above example the environment changed, and this is manifestly favorable to natural selection because it increases the likelihood that profitable variations will occur; without profitable variations, natural selection can do nothing. Not that extreme variability is necessary; humans can produce great results by adding up mere individual differences. Nature can do the same, only far more easily, because she has incomparably more time at her disposal. Moreover, great physical changes, as of climate or unusual isolation to check immigration, are not necessary for the creation of new unoccupied niches for natural selection to fill by modifying and improving some of the varying inhabitants. Because all the inhabitants of a region struggle together, the forces between them being finely balanced, extremely small variations in the structure or habits of one inhabitant would often give it an advantage over others; further variations confer still further advantages. There is no region where all the native inhabitants are so perfectly adapted to one another and their physical environment that none could be improved. In all regions, native species have been conquered to some extent by invasive organisms, which shows that the natives could have been modified in ways that would have helped them to resist the intruders better.

Given that humans produce such great results through methodical and unconscious selection, is there anything nature can't effect? Humans can select based only on external appearances, but Nature cares nothing for appearances, except insofar as they are useful. She can act on every internal organ and every shade of constitutional difference: on the whole machinery of life. Humans select only for their own good; Nature, only for that of the being she tends. Every selected characteristic is exercised by her, and the organism is thus placed under well-suited conditions of life. Humans keep organisms from many different climates in the same place, seldom exercising each selected characteristic uniquely and fittingly: feeding long-beaked and short-beaked pigeons on the same food, not exercising long-backed or long-legged quadrupeds in any particular way, exposing sheep with long and short wool to the same climate, preventing the most vigorous males from competing for the females, failing to unyieldingly destroy all inferior animals, and protecting all as much as possible during each season. Humans often begin selection with some half-monstrous form, or at least with some eye-catching prominence or plainly useful trait. But in the wild the slightest structural or constitutional difference may shift the balance of the struggle for life and so be preserved. How fleeting are human wishes and efforts! How little their time and how poor their products when compared with those accumulated by nature during whole geological periods! No wonder Nature's productions are "truer" in character, infinitely better adapted to complex environments, and plainly bear the mark of far higher workmanship.

Natural selection constantly scrutinizes every variation, rejecting the bad while preserving and accumulating the good; whenever and wherever opportunity affords, nature works silently and unnoticed to improve every organism with respect to its environment. We see nothing of these slow changes in progress until the hand of time has marked the long lapse of ages. Even then our view into long-past geological ages is so imperfect that we can observe only that life forms are different now from what they were in the past.

Natural selection can act only through and for the good of each organism, and in this way it also acts on characteristics that seem insignificant to us. When we see leaf-eating insects colored green and bark feeders a mottled gray, or the alpine ptarmigan white in winter, the red grouse the color of heather, and the black grouse the color of peaty earth, the obvious conclusion is that tints protect these insects and birds from danger. Grouse would increase to countless numbers if they were not subject to destruction at some period of their lives; they are known to suffer largely from birds of prey; hawks use eyesight to find prey (in continental Europe people are warned not to keep white pigeons, because these are likely to be destroyed). Thus natural selection is extremely effective at giving – and maintaining – the appropriate color of each kind of grouse. And the occasional destruction of an animal with a particular color has a nontrivial effect; it is essential to destroy every lamb with the faintest trace of black in order to maintain a white flock. Botanists consider the down on fruit and the color of flesh negligible characteristics, but the horticulturalist Downing reports that in the United States, smooth-skinned fruits suffer more from a beetle, a curculio; purple plums suffer more from a certain disease than yellow plums; and another disease attacks peaches with yellow flesh more than it does others. If with all the aids of art these minor differences have an impact when cultivating varieties, then surely in the wild, where the trees struggle with other trees and a host of adversaries, such differences would settle which variety – smooth or downy, yellow or purple – will succeed.

In considering characteristics that, as far as our ignorance permits us to judge, seem unimportant, remember that climate, food, and other external factors probably produce some minor and direct effect. It is, however, much more important to remember that there are many unknown rules of correlated growth that cause often unexpected modifications in one part when variations are accumulated in another by natural selection.

Under domestication, variations that appear at a particular period of life tend to appear in the offspring at the same period. Examples include variations in the seeds of many culinary and agricultural plants; in the cocoons and caterpillars of silkworms; in the eggs of poultry and the down of their chicks; and in the horns of nearly adult sheep and cattle. The same is true in the wild, which means that natural selection can accumulate profitable variations at any age, to be inherited and exhibited at a corresponding age in the offspring. If it is profitable for a plant to have its seeds disseminated more and more widely by the wind, then natural selection can effect this as easily as the cotton planter selecting for increased down in the pods of cotton trees. Natural selection can modify an insect larva to many contingencies that are totally different from those faced by the mature insect, but these modifications are also likely to influence the adult through correlation. This may mean that among those insects that live for only a few hours as an adult and never feed, a large part of their structure is the correlated result of successive changes in the structure of their larvae. Conversely, modifications in the adult probably influence the larva. But in all cases natural selection ensures that modifications with correlated effects at another period of life will not then be detrimental; otherwise they would cause the species to become extinct.

Natural selection modifies the young relative to the parent and the parent relative to the young. In social animals it adapts each individual for the benefit of the community if both individual and community profit by the selected change. What natural selection cannot do is modify one species, without giving it some advantage, solely for the good of another species. Some works on natural history state otherwise, but I cannot find a single case that stands up to investigation. A structure that is used only once in an animal's life but is very important to it can be modified to any extent by natural selection. For example, the great jaws of some insects are used exclusively to break open the cocoon, and nestling birds have hard-tipped beaks to break open the egg. It is said of the best short-beak tumblers that more of them perish in the egg than are able to get out, so breeders have to assist hatching. But if nature had to make the beak of the adult pigeon very short for the bird's own advantage, modification would be slow, and at the same time there would be rigorous selection of the young within the egg: those with the hardest and most powerful beaks would survive and those with weak beaks would perish. Alternatively, delicate and more easily broken shells might be selected for, because the thickness of the eggshell varies just like any other structure.

Under domestication, a peculiarity can develop and become hereditarily attached to one sex but not the other. The same is probably true in the wild, which means that natural selection can modify one sex in its relations with the other or generate different habits in the two sexes (as in some insects). This leads me to write a few words about what I call sexual selection, which depends not on a struggle for existence, but on a struggle between males for the possession of females. The result is not the death of unsuccessful competitors, but few or no offspring; therefore sexual selection is less rigorous than natural selection. Usually the males best fitted to their place in nature leave the most progeny, but in many cases victory depends instead on males having specialized weapons. A hornless stag or spurless cock is unlikely to leave offspring. By always allowing the victor to breed, sexual selection can give indomitable courage, long spurs, and strength to the wing (so as to strike with the spurred leg), just as well as the brutal cockfighter who knows he can improve a breed by carefully selecting the best cocks. I don't know how low on the scale of nature this rule of battle descends. Male alligators have been described as fighting, bellowing, and whirling around for the possession of females; male salmon have been seen fighting all day; and male stag beetles often bear wounds from the huge mandibles of other males. The war is perhaps fiercest between males of polygamous animals, and these most often sport special weapons. Males of carnivorous animals are already well armed, but sexual selection can provide them and others with special means of defense like the mane of the lion, shoulder pad of the boar, or the hooked jaw of the salmon. After all, the shield may be as important for victory as the sword or spear.

Among birds the contest is often more peaceful. Ornithologists agree that males of many species spar in song to attract females; nonetheless, the rivalry is often intense. Some birds, like the rock thrush of Guiana and birds of paradise, congregate, with successive males displaying their gorgeous plumage and performing strange antics in front of watching females, who eventually choose the most attractive partner. Observations of birds in confinement show that individuals have particular likes and dislikes. Sir R. Heron has described how one pied peacock was eminently attractive to all of his hens. It may seem childish to attribute any effect to such apparently weak means, and I cannot go into the details necessary to support this idea, but if humans can, in a short time, endow bantams with characteristics in accordance with a human standard of beauty, then I cannot see why female birds should be incapable of endowing males with their conception of beauty by selecting, over thousands of generations, the most melodious and attractive males. I strongly suspect that sexual selection can explain some well-known rules concerning the adult plumage of male and female birds in comparison to the plumage of the young. I suspect that plumage is chiefly selected at the breeding age or breeding season, with the resulting modifications inherited to become manifest at the corresponding age or season.

Thus, when males and females of a species have the same general habits but differ in structure, color, or ornament, I believe the differences are caused mainly by sexual selection. That is, in a succession of generations, individual males had some slight advantage over other males in weaponry, defense, or charm that was passed on to male offspring. I do not want to attribute all sexual differences to this agency. Certain peculiarities in male domestic animals, such as the carrier's wattle or the hornlike protuberances in the cocks of certain fowls, cannot be useful in battle or attractive to females. There are analogous cases in the wild: the tuft of hair on the breast of the turkey cock, for example. (Indeed, had the tuft appeared under domestication it would have been called a monstrosity.)

To illustrate how I believe natural selection works, I will give a few hypothetical examples. The wolf preys on various animals and catches some by craft, some by strength, and some by fleetness. If some environmental change were to result in more of the fastest prey (say, deer) or fewer other prey during the season when the wolf is most pressed for food, then the swiftest and slimmest wolves would have the best chance of surviving and would thus be selected. (This assumes that the wolves always retain enough strength to master other prey when necessary.) I can see no more reason to doubt this than the human ability to improve the fleetness of the greyhound by careful and methodical selection, or by unconscious selection from each breeder keeping the best dogs without any thought of modifying the breed.

Even without any change in the relative numbers of the wolf's prey, a cub might be born with an innate tendency to pursue certain kinds of animals. This is not improbable, given that the natural tendencies of domestic animals vary greatly.[1] A wolf with any slight innate beneficial change in habit or structure would have the best chance of surviving and leaving offspring. Some of its young would be likely to inherit the new

---

1. Some cats prefer rats, others mice. According to Mr. St. John, one brings home winged game, another hares or rabbits, and another hunts on marshy ground and nightly catches woodcocks or snipes. The tendency to catch rats rather than mice is known to be inherited.

habit or structure, and by the repetition of this process a new variety would be formed that may either supplant or coexist with the parent form. Or wolves inhabiting a mountainous region and wolves inhabiting lowlands would naturally hunt different prey (whichever the particular environment supplied); continued preservation of individuals best suited to the two sites would result in the slow formation of two varieties. (These varieties would cross and blend wherever they met – I will return to this subject.) According to Mr. Pierce, two wolf varieties inhabit the Catskill Mountains of the United States: one has a light greyhound-like form and pursues deer; the other is bulkier, with shorter legs, and more frequently attacks the shepherd's flocks.

Let's consider a more complex case. Certain plants secrete a sweet juice through glands at the base of the stipules in some legumes and the back of the leaf in the common laurel, apparently to remove something harmful from their sap. Insects seek this juice greedily, even though there is very little of it. Now, suppose a flower secretes a little sweet juice or nectar from the inner bases of the petals. Insects seeking this nectar would get dusted with pollen and transport it to the stigma of another flower. Distinct individuals of the same species would thus get crossed. Because crossing produces more vigorous seedlings, these would have the best chance of flourishing and surviving. Some of these seedlings would likely inherit the nectar-secreting ability. Those with the largest glands or nectaries would be most frequented by insects and most crossed, and so in the long run would gain the upper hand. Given the particular insects that visit them, flowers with the stamens and pistils best placed to facilitate the transport of pollen from flower to flower would also be favored and selected. We could also consider insects visiting flowers for pollen instead of nectar. Because the purpose of pollen is fertilization, its destruction would appear to be a loss to the plant, but if a little pollen is carried by the pollen eater from flower to flower, at first occasionally and later habitually, the gain to the plant is huge, even if nine-tenths of the pollen is destroyed. Plants producing more and more pollen with larger and larger anthers would thus be selected.

The natural selection of more and more attractive flowers will render a plant highly tempting to insects. These will then – unintentionally on their part – carry pollen from flower to flower. Although there are many striking examples of insects fulfilling this role, I will give only one. This example also illustrates one step in the separation of sexes in plants. Some holly trees bear only male flowers, with four stamens producing a small quantity of pollen, along with a rudimentary pistil. Other holly trees bear only female flowers. These have a full-sized pistil and four stamens with shriveled anthers devoid of pollen. I found a female tree exactly sixty yards from a male tree. I inspected the stigmas from twenty flowers (all taken from different branches) under a microscope and found pollen grains on all of them – some were profusely covered. For several days the wind had been blowing from the direction of the female tree toward the male tree, so the pollen could not have been transported this way. The weather had been cold and boisterous and therefore unfavorable to bees, and yet every female flower I examined had been fertilized by bees, accidentally dusted with pollen while they searched for nectar.

Returning to the hypothetical case, as soon as the plant had been rendered highly attractive to insects, another process might commence. Naturalists agree that a so-called physiological division of labor is advantageous: a plant benefits from developing stamens alone in one flower or on the whole plant and pistils alone in another flower or another plant. In cultivated plants, the male or female reproductive organs sometimes become somewhat impotent. In the wild, pollen is already transported among flowers, and because sexual division is advantageous under the principle of division of labor, it will be selected until the sexes become completely separated.

Consider the nectar-feeding insects in the hypothetical scenario. Suppose that the plant slowly gaining nectar by natural selection is common and that certain insects depend on the nectar for food. Many observations show how anxious bees are to save time; for example, they have a habit of cutting a hole in the base of a flower to suck nectar when they could enter by the mouth with very little extra trouble. Given this, an accidental deviation in body size or form, or the curvature or length of the proboscis, or some other change that is too minute for our appreciation, might allow a bee or other insect to obtain food more quickly, survive better, and leave descendants that inherit the deviation. The tubes of the corollas of the incarnate and common red clovers do not superficially

appear to differ in length, and yet honeybees can easily suck nectar out
of the first but not the second, which is visited only by bumblebees. Entire fields of red clover offer abundant nectar in vain to the honeybee. It
might be a great advantage to the honeybee if its proboscis were slightly
longer or differently constructed. Conversely, I have found by experiment that the clover's fertility depends on bees visiting and moving parts
of the corolla to push pollen onto the stigmatic surface. Therefore, if
bumblebees became rare, it would be advantageous to the red clover to
develop a shorter or more deeply divided tube to its corolla, thus allowing honeybees to visit its flowers. This explains how a flower and a bee
might simultaneously or sequentially become slowly modified and perfectly adapted to each other by the preservation of individuals presenting
mutually favorable deviations of structure.

   I am aware that this concept of "natural selection" as illustrated by
these hypothetical examples is open to the same criticisms that first
befell Sir Charles Lyell's views on "the modern changes of the earth, as
illustrative of geology." But by now the action of coastal waves is rarely
called "insignificant" when applied to the excavation of gigantic valleys or the formation of long lines of inland cliffs. Natural selection can
act only by the preservation and accumulation of infinitesimally small
inherited modifications, each profitable to the preserved being. Just as
modern geology has almost banished views such as the excavation of
great valleys by a single diluvial wave, so natural selection, if correct, will
banish the belief of continued creation of new organisms or sudden and
great modifications to their structures.

   It is obvious that animals and plants with separate sexes must mate to
reproduce, but this is not obvious for hermaphrodites. They nevertheless do mate – at least occasionally, but sometimes habitually – as first
suggested by Andrew Knight. This is an important point, but I will treat
it only briefly, even though I have material for an extensive discussion.
All vertebrates, all insects, and some other large groups of animals mate
to reproduce. Modern research has diminished the number of supposed
hermaphrodites, and many true hermaphrodites are known to mate.
There are nonetheless many hermaphroditic animals that do not ha-

bitually mate, and the vast majority of plants are hermaphrodites. In these cases is there any reason to suppose that individuals ever mate to reproduce?

I have collected a large set of facts showing that a cross between different varieties or strains of an animal or plant gives vigor and fertility to offspring. Breeders almost universally agree. Moreover, close inbreeding reduces vigor and fertility. These facts alone suggest a law of nature: no organism self-fertilizes for endless generations. Rather, an occasional cross with another individual, even if at very long intervals, is indispensable. (Despite this conclusion, we are utterly ignorant of why it should be so.)[2]

Assuming this is a law of nature, it illuminates many observations that are otherwise inexplicable. For example, exposure to water is unfavorable to the fertilization of flowers, yet many flowers have their anthers and stigmas fully exposed to the weather. But if an occasional cross is essential, it makes sense to allow full freedom for the receipt of pollen from another individual, especially because the flower's own anthers and pistil are usually so close together that self-fertilization seems inevitable. Many plants – such as members of the pea family – enclose their reproductive organs, but in several if not all such plants there is a curious adaptation between the flower and the way in which bees suck the nectar: while feeding they either push the plant's own pollen onto the stigma or deliver pollen from another flower. I have found in experiments published elsewhere that the fertility of such plants is greatly diminished if visits from bees are prevented. It is hardly possible for bees to fly from flower to flower and not transfer pollen, to the great good of the plant. Bees act like a camel-hair pencil: it is enough to just touch the anthers of one flower and then the stigma of another with the same brush to ensure fertilization. Yet bees do not produce a multitude of hybrids between distinct species this way. As Gärtner has shown, if a brush is used to put a plant's own pollen and pollen from another species on the anthers, the first will invariably eliminate any influence from the foreign pollen.

2. [We now know that the purpose of sex is to shuffle genetic information; it yields the "hybrid vigor" noted by Darwin. – D.D.]

When the stamens of a flower spring or slowly and successively move toward the pistil, the purpose seems to be self-fertilization, and it is probably useful to this end. But manipulations by insects are often required to cause the stamens to spring forward, as Kölreuter has shown with the barberry. Curiously, plants of this very genus, with such an apparent capacity for self-fertilization, are almost impossible to raise pure when closely related forms or varieties are planted near one another, because crosses happen so naturally. Many other plants prevent the stigma receiving pollen from its own flower, as reported in the writings of C. C. Sprengel and from my own observations. For example, Lobelia fulgens employs a beautiful and elaborate contrivance by which all the pollen grains are swept out of the conjoined anthers of a flower before the stigma is ready to receive them. This plant is never visited by insects in my garden and so never sets seed, but by manually placing pollen from one flower on the stigma of another I raised many seedlings. Another species of Lobelia growing nearby is visited by bees and seeds freely. C. C. Sprengel has shown, and I can confirm, that in many other cases either the anthers burst before the stigma is ready, or the stigma is ready before the pollen of that flower is ready; these plants have effectively separated sexes and therefore must cross all the time rather than self-pollinate. How strange that the anther and stigma of a flower should be so close together, as if for the very purpose of self-fertilization, but in many cases actually be useless to each other! This is simply explained if occasional crossing with a distinct individual is advantageous or indispensable.

If several varieties of cabbage, radish, onion, or of some other plants are allowed to seed near one another, a large majority of seedlings turn out to be mongrels. For example, I raised 233 cabbage seedlings from plants of different varieties growing near one another, and only 78 were true to their kind (even some of these were not perfectly true). Yet the pistil of a cabbage flower is surrounded not only by its own six stamens but also by those of many other flowers on the same plant. So how, then, are so many of the resultant seedlings mongrelized? I suspect that pollen from a distinct variety overpowers the pollen from the same flower and that this is part of the general law of the advantage of crossing distinct individuals of the same species. (When distinct species are crossed, the rule is reversed and the plant's own pollen overpowers foreign pollen. This subject will be treated in chapter 8.)

In the case of a huge tree with many flowers, a valid objection is that pollen rarely moves from tree to tree but at most from flower to flower on the same tree – and that flowers on the same tree are distinct individuals in a limited sense. Nature provides against this, as most trees bear flowers with separated sexes. If male and female flowers are produced on the same tree, pollen needs to be carried from flower to flower, increasing the chances of occasional crosses between individual trees. In this country I find that separate sexes are more common in all tree orders than in other plants. At my request Dr. Hooker tabulated the trees of New Zealand and Dr. Asa Gray those of the United States, and the results were as I had anticipated. However, Dr. Hooker recently informed me that the rule does not hold in Australia. I have made these few remarks on the sexes of trees simply to draw attention to the subject.

Turning briefly to animals, hermaphrodites on land include land mollusks and earth worms, but these all mate. I have yet to find a single terrestrial animal that fertilizes itself. This necessity for crossing contrasts sharply with terrestrial plants and makes sense considering where animals live and the nature of the fertilizing element. For animals there is nothing analogous to the action of insects or wind to effect cross-fertilization without the mating of two individuals. There are many self-fertilizing aquatic animals, but currents in the water offer an obvious means for occasional crossing. Even after consulting with Professor Huxley – one of the highest authorities on the matter – I have not discovered a single hermaphroditic animal with reproductive organs so completely enclosed that mating with a distinct individual would be physically impossible. Barnacles seemed exceptional for a long time, but by a fortunate chance I could prove that although individual barnacles are self-fertilizing hermaphrodites, they sometimes cross.

It must strike most naturalists as a strange anomaly that among both plants and animals, species in the same family or even genus often agree almost completely in organization, and yet some are hermaphrodites while others have two sexes. However, if all hermaphrodites do in fact occasionally cross, then the functional difference between hermaphrodites and species with separate sexes becomes very small.

Based on these considerations and other facts I have collected, I suspect that the need for occasional crossing is a law of nature for both plants and animals. I am aware that there are many challenging cases, some of which I am investigating. To sum up, crossing between two individuals is necessary for reproduction in many organisms, in many others it occurs perhaps only at long intervals, but in no case can self-fertilization go on forever.

The circumstances favorable to natural selection present an extremely intricate subject. A large amount of heritable and diverse variability is favorable, but mere individual differences suffice for natural selection to work. An important point is that the existence of a large number of individuals increases the chance that a profitable variation will appear in a given period of time, compensating for low levels of variability in individuals. Although Nature grants vast amounts of time for the work of natural selection, she does not grant infinite time. Every organism strives to seize a place in nature, and a species is quickly exterminated if it fails to become modified and improved as its competitors do.

A breeder's methodical selection works toward a definite goal, and free crossing ruins his work. When many breeders have the same standard of perfection and try to use the best animals, they unintentionally improve and modify the breed by unconscious selection, despite frequent crossing with inferior animals. The same occurs in the wild. In a confined region with some niche not exploited to its full potential, natural selection tends to preserve individuals that vary in the "right direction" in order to better fill the unexploited niche. But a large area presents multiple environments, and as natural selection modifies and improves a species in each of these environments, individuals of the same species will cross at the borders. In this case the effects of interbreeding cannot be counterbalanced by natural selection, because in a continuous region, conditions change gradually from one area to another. Intercrossing will most affect animals that mate to reproduce, wander a great deal, and breed slowly. Varieties of such animals (birds, for example) are generally confined to separated regions. Among hermaphrodites that cross only occasionally, and also among animals that mate to reproduce but wander little and can proliferate rapidly, a new and improved variety may form quickly on the spot; interbreeding would take place mainly between individuals of the same new variety. A local variety formed this way may subsequently spread (albeit slowly) to other regions. This is why planters prefer getting seed from a large number of plants belonging to the same variety, thereby reducing the chance of intercrossing with other varieties.

Nonetheless, intercrossing does not severely retard natural selection, even in slow-breeding animals that mate to reproduce. I can bring a considerable catalog of facts to show that varieties of the same animal can remain distinct for a long time within the same area because they haunt different habitats, breed during slightly different seasons, and prefer to pair with their own variety.

Crossing plays an important part in keeping a species or variety true and uniform. It obviously acts more efficiently with animals that mate to reproduce, but I have already shown that occasional crossing takes place with all animals and plants. Even if crosses occur only at long intervals, I am convinced that the resulting offspring are more vigorous and fertile than offspring from long-continued self-fertilization and thus have a better chance of surviving and propagating their kind. And so, in the long run, the influence of even occasional crosses is great. An organism that never crosses with other varieties remains uniform only as long as conditions stay constant, through inheritance and natural selection destroying individuals that vary from the proper type. If the environment changes and the species becomes modified, the altered offspring will be uniform solely because natural selection preserves the same favorable variations in each.

Isolation is also important to the process of natural selection. In a small and isolated area, the environment will generally be uniform, so natural selection will tend to modify all the individuals of a varying species in the same way. In addition, isolation precludes intercrossing with individuals of the same species from a nearby but different environment; it also checks immigration of better-adapted organisms. New niches created after a physical change (climate, land elevation, etc.) thus remain open to the original inhabitants, which struggle and become adapted through constitutional and structural modifications. By checking immigration and therefore competition, isolation provides nascent varieties with time to slowly improve, and this may sometimes be important in the generation of new species. However, a very small isolated area enclosed by barriers or with peculiar physical conditions necessarily supports small populations, greatly retarding the generation of species through natural selection by decreasing the chance that favorable variations will appear.

A small oceanic island exemplifies a "small isolated area." The total number of species on oceanic islands is small (as will be described in the chapter on geographic distribution), but most of these species are endemic, meaning they are produced there and nowhere else. An oceanic island may therefore seem highly favorable to generating species, but to see if this is actually the case, a comparison within equal times is necessary to ascertain if it is more favorable than a large open area like a continent, and this we are incapable of doing.

Indeed, I believe that, on the whole, a large area is more important than isolation for the generation of new species, especially those that will endure and spread widely. A great and open area harbors large populations within which favorable variations are more likely to arise. The environment is also infinitely more complex from the large number of already existing species, and if these become modified and improved, others must improve correspondingly or be exterminated. As a new form becomes improved, it can spread over an open and continuous area and face competition with many others; more new niches form, and the competition to fill them will be more severe. Although large areas may now be continuous due to oscillations of level, many were recently fragmented, so the effects of isolation will have already occurred to a certain extent. I conclude that although small isolated areas have probably been favorable to the generation of new species, modification is more rapid in large areas, and its new forms – already victorious over many competitors – play an important part in the changing history of the organic world by spreading widely and giving rise to new varieties and species.

These observations illuminate some facts that will be alluded to again in the chapters on geographical distribution. For example, organisms of the smaller Australian continent have in the past succumbed and still yield to organisms from the larger Eurasian continent. For the same reason, continental organisms have become naturalized on many islands. On an island the race for life is less severe, so there is less modification and extermination. Perhaps this is why the flora of Madeira, according to Oswald Heer, resembles the extinct Tertiary flora of Europe. Similarly, all freshwater bodies, taken together, are small compared to the oceans or continents, so freshwater organisms experience less competition than organisms elsewhere. New forms develop more slowly and old forms die out more slowly. Consistent with this, freshwater hosts seven genera of ganoid fishes, remnants of a once-prevalent order. Indeed, some of the most anomalous forms now known on the earth are found in freshwater, including the platypus and lungfish, which, like fossils, to a certain extent connect orders that are now widely separated on the scale of nature. These "living fossils" have endured due to the less severe competition within confined areas.

To sum up the circumstances that are favorable and unfavorable to natural selection, as far as the intricacy of the subject permits, I conclude that large continental areas best facilitate the formation of novel terrestrial organisms most likely to endure long and spread widely. Continents probably oscillate in level and will consequently be fragmented for long periods. A continent first exists as a continuous body with many individuals and forms engaged in severe competition. Subsidence creates many islands, each harboring many individuals of a species. Crossing is checked and immigration prevented so that after a physical change the original inhabitants will fill new niches through modifications that become perfected with time. With renewed elevation the islands once again become part of a continent, thereby reinitiating severe competition. The most favored or improved varieties spread, and less improved forms become extinct. Consequently, the proportional numbers of species on the renewed continent change again, providing natural selection with a fair field to further improve the inhabitants and generate still more species.

Natural selection always acts extremely slowly. Its action depends on available niches that may be better occupied by inhabitants undergoing modification, and these often arise due to intrinsically slow physical processes and curtailed immigration of better-adapted forms. But probably the action of natural selection more frequently depends on slow modifications in some inhabitants, disturbing the relationships among many others. Nothing can happen without favorable variations, and variation itself is apparently very slow. Free intercrossing retards the process even further. Many will exclaim that these impediments are sufficient to stop the action of natural selection entirely. I do not think so. However, I do believe that natural selection acts slowly, often over long intervals of time, and generally on only a few inhabitants of a region at the same time. This slow intermittent action of natural selection agrees perfectly with geological records of the rate and manner of how organisms have changed.

If feeble humans can do so much by artificial selection, I see no limit to the amount of change, beauty, and infinite complexity in the adaptations of organisms to one another and to their physical environments that can be generated during the long course of time by nature's power of selection.

Extinction will be discussed more fully in the chapter on geology, but its connection to natural selection necessitates some mention here. Natural selection acts solely by preserving advantageous variations that consequently endure. But the geometric power of proliferation in all organisms means that each region is already saturated with inhabitants; as selected and favored forms increase in number, those that are less favored decrease and become rare. Rarity, as the fossil record reveals, is the precursor to extinction. Any form represented by a few individuals faces the possibility of extinction when the seasons or numbers of its adversaries fluctuate. Furthermore, because new forms are slowly but continually being generated, others must become extinct unless we believe that the total number of forms increases endlessly. Yet the fossil record plainly shows that the number of specific forms has not increased endlessly. It makes sense that they should not have done so, because nature does not provide an infinite number of niches. (Not that we have any way of knowing when a region harbors the maximum number of species. Probably no region is fully stocked yet. There are more plant species crowded together at the Cape of Good Hope than any other place in the world, yet some foreign plants have become naturalized without any native species going extinct.)

Furthermore, species with the most individuals have the best chance of producing favorable variations within a given period of time. Facts given in chapter 2 provide evidence for this: common species present the greatest number of recorded varieties (i.e., incipient species). Rare species are modified or improved more slowly and will consequently lose the race for life to modified descendants of more common species.

From these considerations it inevitably follows that as natural selection forms new species, others become gradually rarer and rarer and ultimately extinct. Forms in the closest competition with those undergoing modification and improvement naturally suffer the most. As explained in the previous chapter, closely related forms (varieties of the same species and species of the same genus or related genera) generally engage in the most severe competition, because they have nearly the same structures, constitutions, and habits. Consequently, a new variety or species generally presses hardest on its kindred, ushering them to extinction. The same process of extermination accompanies domestication as breeders select improved forms. There are many curious cases of new varieties of flowers and new breeds of cattle, sheep, and other animals replacing older inferior kinds. The ancient black cattle of Yorkshire were historically displaced by long-horns, and these, in the words of Youatt, "were swept away by the short-horns as if by some murderous pestilence."

The principle of divergence of character is important to my theory and explains several observations. Even strongly marked varieties having somewhat the character of a species (as witnessed by the hopeless doubts in ranking them) differ less from one another than distinct species differ from one another. Varieties are species in the process of formation – incipient species, according to my theory. How does the lesser difference between varieties burgeon to the greater difference between species? That this *does* happen can be inferred by observing the innumerable species in nature that present well-marked differences, whereas varieties, their supposed prototypes, present slight and ill-defined differences. Mere chance might cause one variety to differ in some characteristic from its parents, and the offspring of this variety may differ even more in the same characteristic, but this process alone does not account for the large differences between varieties of the same species on the one hand and species of the same genus on the other.

As is always my practice, I seek illumination of this subject from domestic organisms, where there is something analogous. A breeder notices a pigeon with a slightly shorter beak while another notices a pigeon with a slightly longer beak. On the principle that breeders do not and will not admire an average standard but like extremes, they go on to choose and breed birds with shorter and shorter or longer and longer beaks, as has been done with tumbler pigeons. Similarly, at an earlier time one person preferred swifter horses while another preferred stronger and bulkier horses. The initial differences would have been small, but continuous selection for swifter horses by some breeders and stronger ones by others augmented the differences with time and resulted in sub-breeds. With the lapse of centuries, these became distinct and established breeds. Inferior animals with intermediate characteristics (neither swift nor strong) were neglected and eliminated as the differences slowly became greater. This is an example of divergence as it occurs in domestication: barely perceptible initial differences increase steadily, and the breeds diverge from one another and the common parent.

But how does this translate to the wild? The more diversified in structure, constitution, and habit the descendants of a single species are, the more efficiently they will seize the widely diversified niches of nature, thereby increasing in numbers.

Animals with simple habits are a clear example. Take the case of a carnivore that has reached its maximum average population size in a region. If the region does not undergo any change, the only way its power to increase can be unleashed is if varying descendants seize niches currently occupied by other animals. Some of them might be enabled to feed on new prey or become less carnivorous, or haunt new habitats and climb trees or frequent water. The more diversified the descendants, the more niches they can occupy. What applies to one animal applies through all time to all animals, assuming they vary; otherwise natural selection can do nothing. The same is true with plants. Experiments have shown that a plot of land sown with grasses of several genera yields more dry plant material by weight than a similar plot sown with one grass species. Sowing equal plots with several mixed wheat varieties as opposed to just one gives the same result. If a grass species varies continuously, and those varieties are selected that differ from one another in the same way that grass species and genera differ from each other, then a large number of individuals from this species (modified descendants included) will survive. Each grass variety and species releases countless seeds annually in striving to increase its numbers, so the most distinct varieties have the best chance of supplanting less distinct varieties in the course of thousands of generations. Varieties, when rendered very distinct from one another, take the rank of species.

Many natural circumstances show that the greatest amount of life can be supported by great diversification of structure. Extremely small areas, especially those open to immigration, with intense contests between individuals, harbor great diversity. For example, I found that a three-by-four-foot piece of turf that had been exposed to the same conditions for many years supported twenty plant species belonging to eighteen genera and eight orders! The same diversity typifies plants and insects on small and uniform islets or in small freshwater ponds. Farmers produce the most food by rotating plants belonging to distantly related orders; nature uses a sort of "simultaneous rotation." Most of the animals and plants living near a small piece of ground could live on it and strive to do so (assuming the ground is not somehow peculiar). Because of the advantages conferred by diversified structure and the accompanying differences in habit and constitution, wherever the inhabitants are in closest competition, jostling one another most closely, they will generally belong to different genera and orders.

The same principle operates when humans introduce plants to foreign lands. It might be expected that plants that successfully become naturalized will be closely related to indigenous plants because these are commonly considered to have been specially created and adapted to their own region. It might also be expected that naturalized plants will belong to a few groups that are more especially adapted to certain habitats in their new homes. What actually happens is very different. Alph. de Candolle has remarked in his great work that naturalized plants add far more new genera than new species in proportion to the number of native genera and species. To give an example, the last edition of Dr. Asa Gray's *Manual of the Flora of the Northern United States* enumerates 260 naturalized plants belonging to 162 genera, which shows that these naturalized plants are highly diversified. Moreover, they differ greatly from the indigenous plants, because at least 100 of the 162 introduced genera are not native, constituting a relatively large addition to the genera of the United States.

Invasive organisms that have successfully struggled with native species and become naturalized provide a rough idea of how the natives could have been modified in order to have gained an advantage over other natives; diversification of structure, amounting to new generic differences, would have been profitable to them.

The advantage of diversity among inhabitants of one region is the same as the advantage of physiological division of labor among the organs in a body (a subject elucidated by Milne Edwards). Physiologists agree that a stomach draws the most nourishment by being adapted to digest only vegetable matter or only meat. Thus, the more widely and perfectly organisms are diversified, the more individuals a region can support. A set of animals with modestly diversified organizations cannot compete with a set that is more extensively diversified in structure. As Mr. Waterhouse and others have remarked, Australian marsupials are divided into groups differing only slightly from one another; it seems unlikely that these animals could compete with our carnivores, ruminants, rodent mammals, and other well-differentiated orders. Australian mammals are in an early and incomplete stage of the process of diversification.

The previous discussion could have been amplified, but we can assume that the modified descendants of a species succeed better as they become more diversified and thus enabled to encroach on niches occupied by other organisms. How does this principle – that great benefit derives from divergence of character – function when combined with natural selection and extinction?

The accompanying diagram will aid in understanding this perplexing subject. Let (A) through (K) represent species from a large genus in their native region. They resemble one another to varying degrees – as in the wild. The degree of resemblance is represented by unequal distances between letters on the diagram. I specify a large genus because, as described in chapter 2, species from large genera tend to vary more than those from small genera, and the varying species of large genera have more varieties. Also, common and widely diffused species vary more

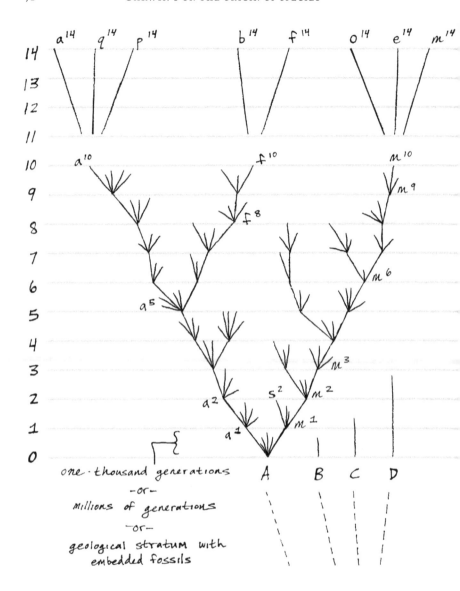

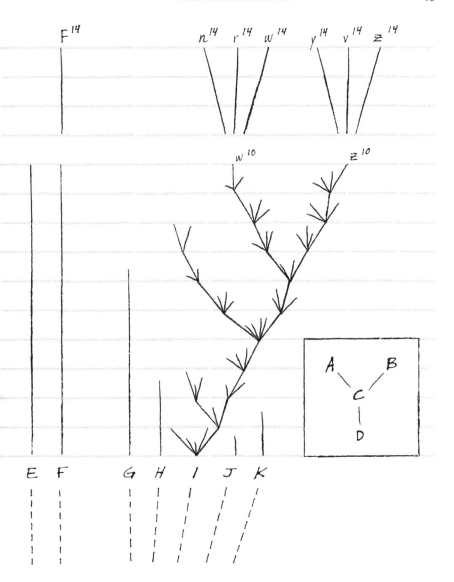

than rare species with restricted ranges. Let (A) be a common, widely diffused, varying species belonging to a large genus residing in its native region. The little fan of unequal diverging lines emanating from (A) represents its varying descendants. The variations are imagined to be extremely small but diverse, not appearing simultaneously but rather after long intervals of time and enduring for unequal periods. Only profitable variations are preserved – that is, naturally selected. This is where the importance of the benefit derived from divergence of character comes in: the most divergent variations (represented by the outer fanning lines) are preserved and accumulated by natural selection. When a line reaches a horizontal mark, sufficient variation has been accumulated to form a fairly distinct variety of the sort worthy of record in a systematic work.

Let an interval between two horizontal lines represent a thousand generations, although ten thousand generations would be more appropriate. After a thousand generations, species (A) has given rise to the two established varieties a^1 and m^1. These two varieties continue to face the same conditions that made their parents variable, and because the tendency to variability is itself hereditary, they will continue to vary, generally in the same way as their parents. Moreover, the two varieties are only slightly modified forms and inherit the advantages that made their parent species (A) more numerous than other inhabitants of the region. They also maintain the more general advantages of the genus that (A) belongs to: a large genus in its native region. And as I have shown, these circumstances are favorable to the production of new varieties.

Assuming a^1 and m^1 do continue to vary, as predicted, their most divergent variations will generally be preserved during the next thousand generations. At this point, a^1 produces variety a^2, which differs more from (A) than a^1 differs from (A) because of the principle of divergence. Variety m^1 produces varieties m^2 and s^2, which differ from each other and, more considerably, from (A). This process could be continued for any length of time. After each thousand generations, some varieties produce only a single variety, some produce two or three (each in a more modified condition), and some fail to produce any varieties. The modified descendants of the common parent (A) generally continue increasing in number and diverging in character. The diagram represents the process up to the ten-thousandth generation and in condensed form up to the fourteen-thousandth generation.

The process never progresses with the regularity depicted by the diagram, although it is rendered somewhat irregular, and the most divergent varieties don't always prevail and multiply. A medium form may endure for a long time and may or may not produce modified descendants. Natural selection always acts according to the nature of the places that are either unoccupied or imperfectly occupied by other organisms, and this depends on infinitely complex relationships. But in general, the more structurally diversified a species' descendants, the more niches they will be able to seize and the more their modified progeny will increase. The diagram depicts lines of succession broken at regular intervals by numbered letters denoting distinct varieties, but these are arbitrary and could have been inserted anywhere that considerable divergent variation has accumulated.

The modified descendants of a common and widely diffused species belonging to a large genus have the same advantages that made their parent successful, so in addition to diverging in character they will multiply in number. This is represented in the diagram by several branches proceeding from (A). Modified offspring from later and more highly improved branches will probably replace and destroy earlier and less improved branches; in the diagram, some of the lower branches do not reach the upper horizontal lines. In some cases modification will be restricted to a single line of descent and the number of descendant varieties will not increase, although the amount of divergent modification may increase with successive generations. This scenario can be imagined by removing all lines proceeding from (A) except the one linking a^1 to a^{10}. For example, the English racehorse and English pointer have apparently gone on diverging from their original stocks without either having given off fresh branches.

After ten thousand generations, species (A) has produced forms a^{10}, f^{10}, and m^{10}, which have come to largely, but perhaps unequally, differ from one another and from their common parent. If the amount of change between each horizontal line is very small, then a^{10}, f^{10}, and m^{10} may still be only distinct varieties, or they may have arrived at the indefinite category of "sub-species." Alternatively, by letting the intervals represent more extensive modification, a^{10}, f^{10}, and m^{10} come to represent distinct species! And thus the diagram illustrates the steps by which small differences distinguishing varieties are increased into the larger differences distinguishing species. By extending the process for a greater number of generations, as condensed and simplified in the diagram, we get eight species – a^{14} through m^{14} – all descended from (A). This, I believe, is how species are multiplied and genera are formed.

In a large genus, more than one species is likely to vary. In the diagram I designate a second species (I) that after ten thousand generations produces two distinct varieties or two distinct species, w^{10} and z^{10}, depending on the amount of change represented by the horizontal lines. After fourteen thousand generations, six new species, n^{14} through z^{14}, are produced. In each genus the species that are already very different generally produce the greatest number of modified descendants and have the best chance of filling new and very different niches. This is why I chose the extreme species (A) and the nearly extreme species (I) as those that are largely varied and give rise to new varieties and species. The other nine species of the genus, marked with capital letters, transmit unaltered descendants.

Extinction also plays an important role during the process of modification. In a fully stocked region, natural selection acts by giving the selected form some advantage over the other inhabitants, so in each stage of descent the improved descendants of a species tend to supplant and exterminate their predecessors and parents. Recall that competition is generally most severe between forms that are closely related in habit, constitution, and structure. Entire collateral lines of descent might be conquered by later improved lines of descent. However, if the modified descendant of a species emigrates or quickly adapts to some new habitat so that parent species and descendant are not in competition, both may continue to exist.

If the diagram depicts considerable modification, then species (A) and early varieties are extinct, having been replaced by eight new species, a^{14} to m^{14}, and (I) having been replaced by six new species, n^{14} to z^{14}.

And we can go further than this. Remember that the original species in the genus resemble one another unequally, as is often the case in the wild. Species (A) is more closely related to (B), (C), and (D) than the other species, while (I) is more closely related to (G), (H), (J), and (K). Also, (A) and (I) are common and widely diffused, so they enjoy some advantage over most other species of the genus. Thus, at the fourteen-thousandth generation, their modified descendants probably inherit some of these advantages in addition to having been improved and diversified during descent to fit many related niches. It is therefore very likely that they will have taken the niches of their parents (A) and (I), as well as some of the original species most closely related to their parents, thereby exterminating them. This means that very few of the original species transmit descendants to the fourteen-thousandth generation. Let only species (F), one of the two species least related to the other original nine, transmit descendants to the fourteen-thousandth generation.

The diagram depicts fifteen new species descended from the original eleven. The difference between species a^{14} and z^{14} is much greater than the difference between the most contrasting of the original eleven species because of natural selection's divergent tendency. Moreover, the new species will be related to one another in very different ways. Of (A)'s eight descendants, a^{14}, q^{14}, and p^{14} are closely related from having recently branched off from a^{10}; b^{14} and f^{14} are distinct from a^{14}, q^{14}, and p^{14}, having diverged from a^5. In addition, o^{14}, e^{14}, and m^{14} will be closely related, but because they branched off early in the process of modification, they will be very different from the other five species and may constitute a subgenus or even a distinct genus.

The six descendants of species (I) form two subgenera or even genera. But because at the start (I) differed extensively from (A), and because of inheritance, (I)'s six descendants also differ extensively from (A)'s eight descendants, especially because they have been diverging in different directions. Importantly, the intermediate species that connected (A) and (I) have all become extinct, with the exception of (F), leaving no descendants. Therefore, the eight new species descended from (A) and the six new species descended from (I) have to be ranked as distinct genera, or even distinct subfamilies.

I believe this is how descent with modification produces two or more genera from two or more species of the same genus, the two or more parent species having descended from a single species of an earlier genus.

The diagram indicates this with broken lines beneath the capital letters, converging downward to a single point representing the common ancestor of the several new genera and subgenera.

The new species F^{14} at the top of the diagram is interesting; it has diverged little and is mostly or completely like (F), so its affinities to the other fourteen new species will be curious and indirect. F^{14} has descended from a form that was intermediate between the two now extinct and unknown parent species (A) and (I), so it is intermediate between their two descendant groups. However, these two groups have been diverging from the types of their parents, so F^{14} is not directly intermediate between them but intermediate between the types of the two groups. (Every naturalist can readily bring an example to mind.)

So far we have let each horizontal interval on the diagram represent a thousand generations, but each may represent a million or hundred million generations, or they may represent successive geological strata complete with extinct remains. I will return to this subject in the chapter on geology and show that the diagram illuminates the affinities between fossils. Extinct organisms usually belong to the same orders, families, or genera as living organisms, and yet they are often intermediate between existing groups, which makes sense considering that the extinct species lived during very ancient epochs when the lines of descent had diverged less.

There is no reason to limit the process of modification to the formation of genera alone. If each successive set of diverging lines represents great change, then a^{14} to p^{14}, b^{14} to f^{14}, and o^{14} to m^{14} form three distinct genera descended from (A), and (I)'s descendants form a further two distinct genera. These two sets from two different ancestors differ due to divergence, and they form separate families, or even orders, depending on how much divergent modification the diagram represents. The two new families or orders have descended from two species of the original genus, which in turn have descended from a still more ancient and unknown genus.

As already discussed, within each region, the species of a large genus will most often present varieties or incipient species. This might have been expected. Natural selection acts through the advantages of some forms over others in the struggle for existence, so it acts mostly on those that already possess some advantage. A large group indicates that its species have inherited some advantages from a common ancestor. Thus the contest for producing new and modified descendants takes place mainly between larger groups struggling to proliferate: one large group slowly conquers another and reduces its numbers, as well as its chances for further variation and improvement. Within one large group, later and more highly perfected subgroups tend to supplant and destroy earlier and less perfected subgroups, because they have branched out and seized many new niches. Small and broken groups and subgroups tend to disappear. In the future, then, large and currently successful groups that have suffered the least extinction are likely to continue to increase for a long period of time. But no one can predict which groups will ultimately prevail; after all, many formerly well-developed groups have gone extinct. Looking into the more remote future, a multitude of smaller groups will go extinct – leaving no modified descendants – owing to the steady increase of larger groups. Of the species living at any one period, very few will transmit descendants far into the future. (I will return to this subject in the chapter on classification.) That few ancient species have transmitted descendants and that all the descendants of the same species make a class explains why there are few classes in the main divisions of the animal and plant kingdoms. Although few of the most ancient species may now have living and modified descendants, the earth may have been populated by as many genera, families, orders, and classes in the most remote geological period as it is now.

To sum up, during the long course of ages and under varying conditions, organisms have varied in their organizations. The high geometrical ratio of increase in each species causes a severe struggle for life at some age, season, or year. It is advantageous for organisms to be diverse in structure, constitution, and habit, because the infinite complexity of their relationships to one another and to the environment demands it. Given all this, it would be extraordinary if variations useful to an organism did *not* occur. Individuals privileged with such variation have the best chance of surviving the struggle for life, and because of inheritance they tend to leave similarly privileged offspring. I call this principle of preservation "natural selection." Inherited qualities of a particular period of life appear at the corresponding period in offspring, so natural selection can modify the egg, seed, or young as easily as the adult. Among many animals, sexual selection supplements natural selection by assuring that the most vigorous and best-adapted males leave the most offspring. Sexual selection also endows males with characteristics that are useful to them alone in their struggles with other males.

Whether natural selection really acts in this way in nature must be judged by the strength and balance of evidence given in the following chapters. I have already shown how it entails extinction, and the major role of extinction in natural history is plainly recorded by geology. Natural selection also leads to divergence, because more organisms can be supported by a given area if they diverge in structure, constitution, and habit; the inhabitants of small areas, or naturalized organisms demonstrate this. The more the descendants of one species diversify – as they become modified and struggle incessantly to proliferate – the better their chances of succeeding in the battle of life. The small differences that distinguish varieties of a species from one another steadily grow to differences that distinguish species of a genus or even distinct genera.

Common, widely diffused, and widely ranging species belonging to large genera vary the most. And these tend to transmit to their modified offspring the superiority that makes them dominant. As just remarked, natural selection leads to divergence and to extinction of life forms that are less improved. These considerations explain the affinities among living things. It is truly wonderful that all organisms throughout all time and space are related to one another in groups subordinate to groups. We see everywhere that varieties of the same species are closely related, species of the same genus are less closely and unequally related (forming sections and subgenera), species of distinct genera are much less closely related, and genera are related in different degrees (forming subfamilies, families, orders, subclasses, and classes). The subordinate groups of a class cannot be ranked single file, but seem clustered around points, and these around other points, and so on in almost endless circles. The view that each species has been independently created cannot explain this phenomenon of classification, but natural selection as illustrated in the diagram, entailing divergence and extinction, does.

The relationships among organisms within a class are sometimes represented by a tree – an apt simile. Green and budding twigs represent existing species, and those produced in former years represent the long succession of extinct species. Growing twigs try to branch out on all sides and overtop and kill the surrounding twigs and branches, just as species and groups of species try to defeat others in the great battle for life. The limbs that are now divided into great branches, and these into smaller and smaller branches, were once themselves budding twigs. This connection between past and present buds may well represent the classification of all living and extinct species in groups subordinate to groups. Only a few of the many twigs that flourished when the tree was just a bush have grown into great branches, bearing many others; very few of the species that lived during long-past geological ages have living, modified descendants. Since the first growth of the tree, many limbs and branches have decayed and broken off, and these represent whole orders, families, and genera with no living representatives, known only in the fossil record. A thin straggling branch occasionally springs from a fork low down on the tree; having been favored by some chance, it is still alive at the summit. So we occasionally see an animal such as the platypus or lungfish thinly connecting two large branches of life. Buds give rise to buds, and the vigorous branch out and surpass many feebler branches. So the Tree of Life fills the crust of the earth with dead and broken limbs and covers the surface with ever spreading beautiful branches.

5

VARIATION

IN EARLIER CHAPTERS I SOMETIMES IMPLIED THAT VARIA-
tions are due to chance. Of course this is completely incorrect, but it
illustrates our ignorance of the causes of variation. Some authors believe
that the reproductive system functions in creating offspring that are
similar to the parents *and* in generating individual differences or slight
structural deviations, but the great frequency of variability and monstrosity under domestication suggests that structural deviations somehow result from the environment endured by parents and their ancestors over several generations. I remark in chapter 1 that the reproductive
system is particularly susceptible to environmental changes. (Proving
this requires a long catalog of facts that cannot fit here.) The varying
and plastic condition of offspring results mainly from the functional
disturbance of the reproductive system in parents, with the reproductive
elements seemingly affected before fertilization.[1] It is unknown why a
disturbed reproductive system should cause this or that part of an organism to vary more than usual. We can nevertheless occasionally catch
hints and recognize that there must be some cause for each structural
deviation, however slight.

How much direct effect climate, food, and other external factors have
on an organism is uncertain – probably very little in animals and more in
plants. Such direct influences could not have produced the many striking and complex structural coadaptations among organisms throughout

---

1. In "sporting" plants, which produce highly variable offspring, the buds are affected (in their earliest condition buds appear essentially the same as ovules).

nature. Climate, food, and so on have some minor influence: E. Forbes confidently asserts that shelled mollusks living at their southern limit, or when living in shallow water, have more brightly colored shells than those of the same species farther north or from greater depths. Gould believes that birds of a given species are more brightly colored under a clear atmosphere than when living on islands or near the coast. Wollaston is convinced that insects' residence near the sea affects their colors. Moquin-Tandon gives a list of otherwise non-fleshy plants that develop fleshy leaves when grown near the sea. Other such cases could be cited.

When varieties of one species range into the habitats of other species, they often acquire some of the natives' characteristics. This agrees with the concept that all species are simply well-defined and permanent varieties, illustrated by the above examples. A person who believes in the individual creation of each species would have to admit that one shell was created with bright colors for a warm sea but that another shell became bright-colored by variation when it ranged into warmer or shallower waters.

When a variation is even slightly useful to its owner, we cannot tell how much of that variation results from cumulative natural selection and how much comes from direct environmental effects. Fur dealers know that animals of a given species have better and thicker fur if they live in a severe climate, but how much of this results from the warmest-clad individuals being preserved over many generations, and how much from the direct action of climate? It does seem that climate has some direct influence on the hair of domestic quadrupeds.

There are examples of a single variety being produced under very different conditions and of different varieties of a single species being produced under the same conditions. This demonstrates how indirectly the environment acts. Naturalists know many examples of species that breed true, or do not vary at all, despite very contrasting climates. Such considerations suggest that little weight be given to direct environmental action. As already remarked, the environment plays an important and indirect role by affecting the reproductive system, thereby inducing variability. Natural selection, then, accumulates profitable variations, however slight, until they become appreciable.

Based on observations described in chapter 1, extensive use of certain body parts in domestic animals strengthens and enlarges them, while disuse diminishes them; such modifications are inherited.[2] There are no standards of comparison in the wild by which to judge the effects of use and disuse, because the parental forms are unknown, but many structures can be explained by the effects of disuse. Professor Owen has remarked that flightless birds are a great anomaly in the wild, yet several such birds do exist. The logger-headed duck of South America can only flap along the surface of the water with its wings, which are in nearly the same condition as those of the domestic Aylesbury duck. Large ground-feeding birds rarely take flight except to escape danger. The nearly wing-less condition of birds inhabiting, or having recently inhabited, oceanic islands that are free of predators has been caused by disuse. The ostrich inhabits continents and is exposed to danger from which it cannot fly away, but it can defend itself by kicking. An early ostrich ancestor may have been like the bustard, and as natural selection increased its size and weight, its legs were used more and its wings less until they became incapable of flight.

Kirby has remarked and I have observed that the front feet of many male dung beetles are often broken off. Not one of seventeen specimens in his collection had even a relic left. *Onites appeles* loses its feet so habitu-ally that it has been described as not having them. They are present in some other genera, but only as rudiments. The *Ateuchus*, sacred beetle of the Egyptians, completely lacks front feet. There isn't sufficient evidence to support the idea that such mutilations can be inherited. So I explain the absence of front feet in *Ateuchus* and their rudimentary condition in some other genera by the long-term effects of disuse in their ancestors. The feet are almost always missing in dung beetles, so they must be lost early in life and therefore cannot be much used.[3]

In some cases it might be easy to attribute a structural modification to disuse even though it has resulted from natural selection. Mr. Wol-laston has discovered that of the 550 beetle species inhabiting Madeira, 200 have such deficient wings that they cannot fly and at least 23 of 29 endemic genera have all their species in this condition! In many parts

2. [We now know that so-called acquired traits are *not* heritable. – D.D.]
3. [Use and disuse as a source of heritible variation is a now-defunct theory. – D.D.]

of the world, beetles are frequently blown to sea and perish; as observed by Mr. Wollaston, beetles on Madeira lie concealed until the wind lulls and the sun shines; the proportion of wingless beetles is greater on the exposed Dezertas than on Madeira; and Mr. Wollaston asserts that Madeira lacks otherwise very common groups of beetles with habits requiring frequent flight. These observations suggest that the winglessness of many Madeiran beetles has resulted mainly from natural selection, probably combined with disuse. During thousands of generations, individual beetles that flew least because of slightly less developed wings or indolence had the best chance of surviving, while frequent fliers were more often blown out to sea and destroyed.

Insects on Madeira that are not ground feeders and must use flight to gain sustenance (like the flower-feeding Coleoptera and Lepidoptera) do not have reduced wings, and even enlarged wings, as suspected by Mr. Wollaston. This is compatible with natural selection. Natural selection would enlarge or reduce the wings of a newly arrived insect depending on whether more individuals would survive by successfully battling the winds or by rarely or never flying. With mariners shipwrecked near a coast, it would be better if bad swimmers could not swim at all and stayed on the wreck and if good swimmers could swim farther still.

The eyes of moles and some burrowing animals are rudimentary and, in some cases, covered by skin and fur. This state is probably due to gradual reduction from disuse, possibly aided by natural selection. A burrowing rodent of South America called the tuco-tuco is even more subterranean than the mole, and I was assured by a Spaniard who often caught them that they are frequently blind. One that I kept alive was definitely blind; on dissection the cause appeared to be inflammation of the inner eyelids. Frequent eye inflammation is harmful to any animal, and eyes are dispensable for those with subterranean habits. So a reduction in eye size accompanied by eyelid adhesion with fur overgrowth might be advantageous, in which case natural selection would constantly aid the effects of disuse.

Several animals belonging to very different classes and inhabiting caves in Styria and Kentucky are blind.[4] Some of the crabs retain the

4. [Styria is a state of southern Austria. There is an extensive system of limestone caves in Kentucky. – D.D.]

stalk for the eye, even though the eye is gone – a telescope stand without its telescope. Although eyes are useless to animals living in the dark, it is difficult to imagine how they might be damaging, so the loss of eyes must result wholly from disuse. The eyes of the blind cave rat are huge, and Professor Silliman thought that one specimen regained some slight power of vision after living in light for several days. Similarly to the Madeira example, in which some insects' wings enlarged and others' became reduced through natural selection aided by use and disuse, with the cave rat natural selection tackled the loss of light and enlarged its eyes, whereas disuse by itself did its work with other cave inhabitants.

Few environments are more uniform than deep limestone caverns under comparable climates, so if the blind animals of European and American caves were separately created, then they should be very similar. However, Schiödte and others have remarked that they are not; cave insects from Europe and America are no more closely related than might be expected from the general resemblance of other inhabitants from the two continents. My view suggests that in each region, animals with ordinary vision migrated from the outside into deeper and deeper recesses of caves over the course of successive generations. Schiödte provides some evidence for the gradation of habit that should be observed if this is true: "animals not far remote from ordinary forms, prepare the transition from light to darkness. Next follow those that are constructed for twilight; and, last of all, those destined for total darkness." By the time an animal reaches the deepest reaches after numberless generations, disuse will have obliterated its eyes and natural selection will have effected other changes to compensate for blindness, like longer antennae or feelers. Even with such modifications, cave animals should exhibit similarities to other (surface) inhabitants of the continent. Professor Dana reports that this is the case with some of the American cave animals, and some European cave insects are closely related to those of the surrounding region. It would be difficult to give a rational explanation for this phenomenon in the context of independent creation. It should be expected that several American and European cave inhabitants are closely related, given the already-known relationships between most other organisms from these two continents. I am not surprised that cave animals are anomalous, like the blind fish mentioned by Agassiz, and, with respect to the reptiles of Europe, the blind *Proteus*. I am surprised only that more wrecks of ancient life have not been preserved in the less severe competition among animals of these dark abodes.

Habits are hereditary in plants, like the period of flowering, amount of rain required for germination, and time of sleep. This leads me to write a few words about acclimatization. Species of the same genus commonly inhabit both very hot and very cold regions, and because all the species within a genus descend from a single parent, acclimatization must occur readily during the long process of descent.[5] Each species is adapted to its native climate; species from an Arctic or even a temperate region cannot endure a tropical climate and vice versa. Succulent plants cannot endure a damp climate. But the degree of adaptation of a species to its native climate is overrated. Whether or not an imported plant will endure the English climate often cannot be predicted, and many organisms brought here from warmer regions survive well. The range of organisms in the wild is limited at least as much by competition from other organisms as by adaptation to a particular climate. Regardless of whether plants generally adapt closely to climate, some of them can partially habituate to different temperatures – that is, acclimatize. Pines and rhododendrons raised in England from seeds collected by Dr. Hooker at different elevations in the Himalaya each resist the cold differently. Similar observations have been made by Mr. Thwaites in Ceylon and by Mr. H. C. Watson of European plant species brought from the Azores to England.[6] In historical times several animal species greatly extended their ranges from warmer to cooler latitudes and vice versa. It is unknown whether these animals were strictly adapted to their original climates (although in ordinary cases we can assume this), nor if they subsequently became acclimatized to their new homes.

Domestic animals were originally chosen by pre-civilized humans because they were useful and bred readily in confinement, not because they were subsequently found amenable to ranging widely. The common but amazing ability of domestic animals to withstand different climates

5. [Do not confuse "acclimatization," which is a biological term for physiological changes (in this case over many generations) in response to the environment, with "acclimate," nor "descent (with modification)" with "descent (down a height differential)." – D.D.]

6. [Ceylon is the former name of modern-day Sri Lanka. – D.D.]

and maintain their fertility under them (a more stringent test) supports the argument that many animals in the wild could easily endure very different climates. This argument cannot be pushed too far, because some domestic animals stem from multiple wild stocks; the blood of tropical wolves, arctic wolves, and wild dogs may mingle in domestic breeds. The rat and mouse are not domestic animals, but they have been transported to many parts of the world and range more widely than any other rodent, living in the cold climate of Faroe in the north, the Falklands in the south, and many islands in hot and dry zones. So adaptation to a specific climate is something grafted onto a wide innate flexibility of constitution that is common to most animals. The ability of humans and domestic animals to endure different climates, or modern elephants and rhinoceroses inhabiting tropical or subtropical climates even though their ancestors endured a glacial climate, are not anomalies but simply examples of a common constitutional flexibility that becomes evident under peculiar circumstances.

It is obscure how much of the acclimatization of a species to a particular climate is due to habit and how much to natural selection of varieties with different innate constitutions. Habit and custom have some influence, based on analogy and on the advice given in agricultural works – even in the ancient *Encyclopedia of China* – to be cautious in transporting animals from one region to another. It is unlikely that humans could have selected so many breeds and sub-breeds with constitutions specifically fitted to their own regions, so the result must be due to habit. Natural selection nevertheless continually preserves individuals best adapted to their native regions. Treatises on many kinds of domestic plants state that certain varieties withstand particular climates better than others; works on fruit trees published in the United States recommend certain varieties for the northern states and others for the southern states, and because most of these varieties originated recently, their constitutional differences cannot be due to habit. The Jerusalem artichoke is never propagated by seed in England, so new varieties have not been produced; because it tastes as tender as ever, some have used its case to argue that acclimatization cannot occur! The kidney bean has also been cited this way. But until someone plants kidney beans early enough so that many are destroyed by frost, and collects seeds from the survivors while preventing accidental crosses, and repeats this for twenty generations, the experiment will not even have been tried. And differences in the constitution of seedling kidney beans do in fact appear; an account has been published about some seedlings being hardier than others.

Habit, use, and disuse have in some cases played a considerable part in the structural and constitutional modification of various organs, but their effects have been combined with and sometimes overshadowed by the natural selection of innate differences.

By "correlated growth" I mean that during the growth and development of an organism, the whole organization is meshed together. When slight variations occur in one part and are accumulated by natural selection, other parts also become modified. This very important subject is imperfectly understood. The most obvious case is that of modifications accumulated for the good of the young or larva affecting adult structure. Similarly, a deformation in the early embryo seriously affects the whole organization of the adult. Homologous body parts that are alike at an early embryonic period seem to vary in a linked way; examples include left and right sides of the body, front and hind limbs, and even jaws and limbs (the lower jaw is thought to be homologous with the limbs). This phenomenon is mastered by natural selection; for example, a family of stags once existed with antlers on only one side, and if this had been particularly useful to the breed, it would have probably been rendered permanent by natural selection.

Homologous parts tend to cohere, as remarked by some authors, and this coherence often occurs in monstrous plants. Homologous structures frequently cohere over the course of normal development, as in the union of petals in the corolla to form a tube. Hard parts seem to affect the form of adjoining soft parts; for example, some authors believe that diversity in the shape of bird pelvises causes the diversity in the shape of their kidneys. Others believe that the shape of a human child's head is influenced by the pressure exerted by the mother's pelvis. According to Schlegel, the body shape of the snake and the way it swallows determine the positions of several important viscera.

The nature of the correlation in correlated growth is frequently obscure. St. Hilaire has asserted that we cannot give any reason as to why certain deformations occur together frequently and others rarely.

Curious examples include, in cats, the correlation of blue eyes with deafness and tortoise-shell color with being female; in pigeons, feathered feet and skin between the outer toes and the presence of more or less down in hatchlings with future plumage color; in the naked Turkish dog, hair with teeth. (Homology probably plays a role in the last case. It is not accidental that the two mammalian orders with the most abnormal skin – namely, whales and the Edentata [armadillos, scaly anteaters, etc.] – also have the most abnormal teeth.)

The difference between the inner and outer flowers of some Compositous and Umbelliferous plants best illustrates the importance of correlation in modifying structures independently of utility, and therefore of natural selection.[7] The central florets and ray florets of, for example, the daisy are different. This difference is often accompanied by the abortion of other flower parts, but in some Compositous plants the seeds are also different in shape and texture; even the ovary with its accessory parts is different, as described by Cassini. Some authors attribute these differences to pressure – and the shape of the seeds in ray florets of some Compositae supports this – but as Dr. Hooker informs me, the Umbelliferous species with the densest heads are not necessarily the ones that most often have differing inner and outer flowers. The development of ray petals could hypothetically draw nourishment from other parts of the flower and cause their abortion, but in some Compositae there is a difference between inner and outer florets without any difference in the corolla. Maybe differences result from unequal nutrient flow to the inner and outer florets; in irregular flowers at least, those near the axis are prone to peloria (i.e., becoming regular). I recently observed an example of this in a striking case of correlation: in some garden geraniums the central flower of the truss often loses its dark patches on the upper two petals accompanied by an abortion of the adjacent nectary. When only one of the two upper petals loses its color, the nectary is only shortened.

Concerning the difference between the corolla of inner and outer flowers of a head, or umbel, C. C. Sprengel's idea that ray florets attract insects may not be as far-fetched as it first seems. Insects facilitate fertil-

7. [The modern equivalents of these obsolete family names are Asteraceae and Apiaceae, respectively. – D.D.]

ization in the Compositous and Umbelliferous orders, and if attracting insects is advantageous, then natural selection may have played a role. However, differences in the external and internal structures of seeds are not always correlated with differences in the flowers. It seems impossible that these differences could be advantageous to a plant, yet in the Umbelliferae they are of such apparent importance that the elder de Candolle founded his divisions of the order upon analogous differences.[8] So structural modifications considered important by systematists may result from unknown rules of correlation and, as far as we can tell, be useless to the species.

Structures common to whole groups of species are often falsely attributed to correlated growth even though they are simply due to inheritance. An ancient ancestor may have acquired some structural modification through natural selection followed by a second independent modification thousands of generations later. These two modifications, inherited by a whole group of descendants with diverse habits, would be thought of as "obviously" correlated in some necessary way. Again, some apparent correlations throughout whole orders are due to natural selection. For example, Alph. de Candolle has remarked that winged seeds are never found in fruits that do not open. The rule can be explained by recognizing that natural selection can only generate winged seeds in fruits that *do* open; individual plants producing seeds that are better fit to wafting farther might have an advantage over those producing seeds that are less fit for dispersal. This process could not possibly take place with fruit that does not open.

At about the same time, the elder Geoffroy and Goethe propounded the rule of compensation or balance of growth, or as Goethe expressed it: "In order to spend on one side, nature is forced to economize on the other side." To a certain extent, this is true with domesticated organisms: if excessive nourishment flows to one part or organ, it rarely flows to another part (at least, not in excess). This is why it is difficult to get a cow to give a lot of milk *and* fatten readily. No single cabbage variety yields abundant nutritious leaves *and* copious oil-bearing seeds. If the seeds

8. According to Tausch, in some cases seeds of outer flowers are orthospermous and seeds of the central flowers are coelospermous.

of a domestic fruit atrophy, the fruit itself gains in size and quality. In domestic poultry, a large tuft of head feathers is generally accompanied by a diminished comb, and a large beard by a diminished wattle. The rule is not universally applicable in nature, but many observers, especially botanists, believe it to be correct. I won't give any examples here, because there is no way to distinguish between the effects of (1) a part becoming greatly developed through natural selection while another adjoining part becomes reduced by either the same process or disuse, and (2) reduced nutrient flow to one part because of excess growth in another adjoining part.

Some purported cases of compensation combined with other observations could be merged under the principle that natural selection continually economizes every part of organization. If a previously useful structure becomes less useful in a new environment, natural selection seizes even slight diminutions in its development, because an individual profits from not wasting nourishment on a useless structure. This is the only way to understand a striking observation I made while examining barnacles (many other examples could be given): a barnacle more or less loses its shell (carapace) when it is parasitic within another barnacle, and therefore protected. This happens with the male *Ibla* and in truly extraordinary fashion with the *Proteolepas*. In all other barnacles the carapace consists of three important and highly developed anterior head segments furnished with great nerves and muscles. But in the parasitic *Proteolepas* the whole anterior part of the head is reduced to a rudiment and attached to the bases of the prehensile antennae. This reduction of a large and complex structure rendered superfluous by the parasitism of *Proteolepas* is a decided advantage to each successive individual of the species. In the struggle for life, each individual *Proteolepas* has a better chance of supporting itself by wasting less nourishment in developing a now-useless structure.

Natural selection always succeeds in reducing and saving any part of the organization as soon as it becomes superfluous, without causing some other part to correspondingly overdevelop. Conversely, natural selection may develop any organ without requiring as compensation the reduction of some adjoining part.

St. Hilaire has remarked that as a rule in both varieties and species, when a part or organ is repeated many times – like snake vertebrae or stamens of polyandrous flowers – the number of repeated units is variable, but it is constant if repeated fewer times. The same author and some botanists have further remarked that multiple parts are liable to structural variation. This "vegetative repetition," to use Professor Owen's expression, is apparently a sign of low organization, related to the general opinion of naturalists that organisms low on the scale of nature are more variable than those that are high. I assume in this case that "lowness" means several parts of organization have not been highly specialized for particular functions, and as long as one part has to perform diversified work, it makes sense that it should remain variable. Natural selection preserves or rejects each little deviation less carefully than when the part has to serve a single special purpose. A knife that has to cut all sorts of things can have almost any shape, but a tool for some specific task better have a particular shape. And recall that natural selection can act on each part of each organism solely through and for its advantage.

Some authors have stated that rudimentary parts are apt to be highly variable, and I agree. I will return to the subject of rudimentary and aborted organs and here mention only that their variability seems to stem from their uselessness. Natural selection has no power to check their structural deviations, so rudimentary parts are left free to the play of various rules of growth, effects of long-term disuse, and reversion.

If a certain part in a species is extraordinarily developed compared with the same part in related species, it tends to be highly variable. A similar remark published by Mr. Waterhouse struck me several years ago. I infer from an observation concerning the arm length of the orangutan that Professor Owen has come to a similar conclusion. It is hopeless to convince anyone of this proposition without giving the long array of facts I have collected and that cannot possibly be introduced here. I can only state my conviction that the rule applies extensively. (I hope I have made due allowance for several potential causes of error.) The rule does not apply to even a very unusual part if it is not unusual in comparison with the same part in closely related species. The bat's wing is abnormal within the mammalian class, but the rule does not apply, because there is a whole group (bats) with wings. It would apply only if one bat species had remarkably developed wings in comparison with other species of the same genus. The rule applies especially to unusual secondary sex characteristics in both males and females (although females exhibit remarkable secondary sex characteristics less often).[9] The rule probably applies so explicitly to secondary sex characteristics because they are highly variable, whether or not they are unusual. The rule is not confined to secondary sex characteristics, as clearly demonstrated by hermaphroditic barnacles.[10] I will give a list of remarkable cases in my future work, but here I name only one that illustrates the rule in its broadest application. The opercular valves of rock barnacles are very important structures and differ very little even among different genera, but in one genus – *Pyrgoma* – these valves are marvelously diverse, sometimes entirely different in shape. The amount of variation in individuals of several of the species is so great that the varieties differ more from one another in the characteristics of these valves than do other species of distinct genera.

Birds within the same region vary remarkably little, so I have particularly attended to them, and the rule seems to hold for this class. However, I cannot ascertain if it applies to plants. This would have shaken my conviction were it not for the great variability of plants making it particularly difficult to compare their relative variability.

A remarkably developed part or organ in any species may fairly be presumed to serve some important role for that species, but it is nevertheless especially liable to variation. Why? There is no explanation if we assume that each species has been independently created with all parts as now observed, but the view that groups of species descend from other species and are modified through natural selection provides some light. If all or part of a domestic animal is neglected and no selection applied, that part (for example, the comb of the Dorking fowl) or the whole breed will no longer be very uniform. The breed is said to have "degenerated."

9. "Unusual secondary sex characteristics" is a term applied by Hunter to traits attached to one sex but not directly involved with the act of reproduction.

10. I am referring specifically to Mr. Waterhouse's remark while investigating this order: the rule almost invariably holds with barnacles.

Rudimentary organs, non-specialized organs, and perhaps polymorphic groups present nearly parallel natural cases, because natural selection has not, or cannot, come into full play; organization is left fluctuating. More interestingly, parts of domestic animals currently undergoing rapid change by continued selection are also especially liable to variation. Consider the prodigious differences between the beaks of the different tumblers, the beak and wattle of the different carriers, the carriage and tail of fantails, and so on; these are the parts English breeders now mainly attend. Breeding even sub-breeds (like the short-faced tumbler) to near perfection is notoriously difficult, and individuals departing from the standard appear frequently. A contest plays out between (1) the tendency to revert to a less modified state and an innate tendency to further all kinds of variability, and (2) the power of steady selection to keep a breed true. Selection prevails in the long run, and we do not expect to fail so far as to breed a coarse common tumbler from a good short-faced breed. As long as selection continues, the structures undergoing modification will vary. And due to unknown causes, variable characteristics produced by human selection sometimes become attached more to one sex than the other (generally the male), as with the wattle of carriers and the enlarged crop of pouters.

Returning to nature, when a part has been extraordinarily developed in comparison to the same part in other species of the genus, then that part has undergone an extraordinary amount of modification since the period when the species branched off from the common ancestor of the genus. This period is seldom very remote, because species rarely endure for more than one geological period. An extraordinary amount of modification implies unusually intense long-term variability that has been accumulated by natural selection for the benefit of the species; as this process had been occurring relatively recently, the part should still be more variable than other parts of the organization that have remained nearly constant for much longer. And I am convinced that this is the case. The contest between reversion/variability and natural selection will eventually cease, and even the most abnormally developed organs may become constant. If an abnormal organ has been transmitted to many modified descendants (as in the case of the bat's wing), it must have existed for an immense period in nearly the same state and come to be no more variable than any other structure. Only in cases where modification has been extraordinarily great and relatively recent should "generative variability" still occur; variability has not yet been fixed by continued selection of individuals varying the right way and to the right extent and continued rejection of those that revert to a less modified former condition.

These principles can be extended. Specific characteristics are more variable than generic characteristics. For example, if in a large genus some species have blue flowers and some have red flowers, then flower color is only a specific characteristic. It would not be surprising if one of the blue species varied into red or vice versa. But if all the species of the genus have blue flowers, then color would be a generic characteristic and its variation unusual. I chose this example because the explanation most naturalists would advance does not apply. They would argue that specific characteristics are more variable than generic characteristics because they distinguish parts of lesser physiological importance than those used to define genera. This explanation is partly and only indirectly true, but I will return to this subject in the chapter on classification. Further evidence to show that specific characteristics are more variable than generic characteristics would be superfluous. But I have repeatedly noticed in works on natural history that when an author mentions with surprise an important part or organ that is constant throughout large groups of species and has differed in closely related species, then it has also been variable in individuals of the same species. This shows that when a characteristic of generic value sinks to a characteristic of specific value, it often becomes variable, though its physiological importance may remain the same. A similar concept applies to monstrosities, or at least St. Hilaire asserts that the more an organ differs in a normal way among species of the same group, the more subject it is to individual anomalies.

If each species had been independently created, why should a structural part that differs from the analogous part in other independently created members of the same genus be more variable than parts that are alike among the species? There is no explanation. However, if species are only fixed varieties, then parts that have varied relatively recently and therefore become distinct can be expected to often *still* vary. Parts in which species of a genus resemble each other are called "generic characteristics," and these are inherited from a common ancestor. (Natural selection is unlikely to have modified multiple species fitted for different habits in exactly the same way.) Generic characteristics are inherited from a remote period when the species first branched off from the common ancestor; they will not have varied and therefore will not have come to differ (not greatly anyway) among member species, so it is unlikely that they should vary in the present day. Parts in which species of a genus differ are called "specific characteristics," and they will have varied and come to differ for each species within the period since it branched off from a common ancestor. It is therefore likely that specific characteristics will still be variable, at least more so than parts of the organization that have remained constant for a very long period.

Secondary sex characteristics are highly variable, and species of the same group vary more widely in these than in other parts. For example, consider the large differences between male gallinaceous birds – which exhibit prominent secondary sex characteristics – with the small differences between females. The original cause of variability in secondary sex characteristics is obscure, but it makes sense that secondary sex characteristics have not been rendered as constant as other parts of organization. They are accumulated by sexual selection, which is less firm than natural selection, because it entails only fewer offspring for males that are less favored, rather than death. Regardless of why secondary sex characteristics vary, they vary extensively, so sexual selection has wide scope for action and readily endows species of the same group with greater differences in secondary sex characteristics than in other parts of structure.

Remarkably, differences in secondary sex characteristics between the two sexes of a species are generally displayed in the same parts that are different between the species of a genus. Members of large groups of beetles commonly have the same number of joints in the tarsi, but the number varies greatly in the Engidae, as Westwood has remarked. The number likewise differs between the two sexes of the same species. Similarly, wing venation in *Tiphia* wasps is important because it is common to large groups, but in certain genera the venation differs among member species and also in the two sexes within a species. This phenomenon has a clear meaning given my view of the subject. All the species of a genus have descended from the same ancestor, as have the two sexes of a species. Consequently, whatever part of the common ancestor, or its early descendants, became variable would be seized by *both* natural selection and sexual selection in order to fit the species into multiple environments and fit the two sexes of the same species to each other – the males and females to different habits, or the males to struggle with one another in the pursuit of females.

I conclude that the following principles are closely connected: specific characteristics vary more than generic characteristics; a part that is extraordinarily developed in a species relative to the same part in other species from the same genus also tends to vary greatly; a part that is common to a whole group of species does not vary greatly even if it is extraordinarily developed; secondary sex characteristics vary greatly and are very different among closely related species; and secondary sex characteristics and regular specific differences are generally exhibited in the same part of the organization. These principles result mainly from (1) species of the same group having descended from a common ancestor from which they inherited much in common; (2) parts that have recently varied extensively being more likely to continue varying than parts that were inherited from a remote period and have not varied; (3) natural selection overcoming the tendency to reversion and furthering variability; (4) sexual selection being less severe than ordinary natural selection; and (5) variations in the same parts having been accumulated by natural selection and sexual selection and thus adapted for specific purposes and secondary sexual purposes.

Distinct species can vary analogously, and a variety of one species often assumes some characteristics of a related species or reverts to some characteristics of an ancestor. These propositions are readily illustrated by considering domesticated varieties. Very distinct pigeon breeds from widely separated regions have subvarieties with reversed feathers on their heads and feet; the original rock pigeon does not possess these characteristics, so they are analogous variations. The fourteen or even sixteen tail feathers frequent in the pouter should be considered a variation representing the normal structure of another breed: the fantail. Undoubtedly, such analogous variation results from pigeon breeds having inherited from a common parent the same constitution and tendency to variation when acted on by similar unknown influences. The enlarged stems, or tubers, of the Swedish turnip and rutabaga present a case of analogous variation among plants. Several botanists rank them as varieties produced by cultivation from a common parent, but if this is *not* true, then they present a case of analogous variation in two so-called distinct species to which a third, the common turnip, can be added. According to the common notion that each species was independently created, the similarity of the enlarged stems of these three plants results not from the true cause of common descent and a consequent tendency to vary similarly, but from three separate yet closely related acts of creation.

The situation is different with pigeons. There is the occasional appearance in all breeds of slate-blue coloration with two black bars on the wings, a white rump, a bar at the end of the tail, and outer feathers externally edged near their bases with white. These are all characteristics of the rock pigeon, so this is a case of reversion and not a case of a new and analogous variation. The marks tend to appear in offspring from crosses between different-colored breeds, and there is nothing in the environment to cause the effect beyond the influence of crossing and inheritance.

Surprisingly, characteristics that disappeared perhaps hundreds of generations ago can reappear. And when one breed is crossed only once with some other breed, offspring retain the ability to revert to the foreign breed for many generations – for twelve or even twenty generations. The foreign proportion of blood in any ancestor after twelve generations would be 1 in 2,048, and yet this is enough to maintain the potential for reversion. If both parents from an otherwise pure breed have lost some characteristic of an ancestor, the potential for regaining the lost characteristic can be transmitted for almost any number of generations. Consider a breed that loses a characteristic that reappears many generations later. The likeliest hypothesis is not that the reverted offspring has suddenly taken after an ancestor from hundreds of generations ago, but that some unknown favorable condition has allowed a latent tendency to reversion, maintained by successive generations, to emerge. For example, on rare occasions the barb pigeon produces a black-barred bird; each generation probably possesses the potential to create birds with this plumage. (Several observations support this.) Inheritance of the *potential* to produce any characteristic is no less likely than inheritance of useless rudimentary organs, which are known to be inherited. Indeed, even the potential to produce a rudiment can be inherited. For example, a rudimentary fifth stamen appears in the common snapdragon so often that the potential to produce it must be inherited.

All species within a genus have descended from a common parent, and so, as expected, the species vary analogously on occasion. A variety of one species might come to resemble another species in some characteristics, because this other species is only an established and permanent variety. But characteristics gained this way are probably unimportant; the presence of important characteristics is governed by natural selection in accordance with the diverse habits of a species, not by the mutual action of environmental influence and a similar inherited constitution. Furthermore, species within a genus might occasionally revert to the same ancestral characteristics. However, we never know the exact characteristics of a group's common ancestor, so these two cases cannot be distinguished. For example, if we did not know that the rock pigeon lacks feathered feet and a turn crown, we could not have discerned whether these characteristics of domestic breeds are reversions or only analogous variations. We could have inferred that blueness is a reversion from the many markings correlated with it (it is unlikely that all would appear together from simple variation), or from its frequent appearance when distinct breeds are crossed. It remains difficult to establish which cases in the wild are reversions and which are new analogous variations. Either way, my theory predicts that species should sometimes assume characteristics already found in some other members of the same group, and this is observed in nature.

A considerable part of the difficulty in recognizing a variable species is that its varieties "mock" other species of the same genus. There are also many cases of intermediates between two other forms, each ranked as either a variety or a species; in this case, one of the two has varied to assume some characteristics of the other, producing the intermediate form, unless these forms were independently created. But the best evidence is provided by important and uniform parts that vary and take on characteristics of the same parts in a related species. I have collected a long list of such remarkable cases, of which I will give only one curious and complex case not involving an important characteristic, but one that occurs in several species of a genus, partly under domestication and partly in the wild. It is an apparent reversion. The ass often has distinct transverse bars on its legs, which are most pronounced in the foal. The shoulder stripe is variable in length and outline and is sometimes double. A white ass (not an albino) has been described as lacking spinal and shoulder stripes, which are sometimes very obscure or missing in dark-colored asses. The koulan of Pallas is said to have been seen with a double shoulder stripe. The hemionus does not have a shoulder stripe, although traces of it occasionally appear (as stated by Mr. Blyth and others), and Colonel Poole informs me that the foals are generally striped on the legs and faintly on the shoulders. The quagga is barred like the zebra over the body but not the legs, although Dr. Gray has figured one specimen with distinct zebra-like bars on the hocks.

I have collected cases of the spinal stripes of distinct horse breeds of all colors in England. Transverse bars on the legs occur on duns,[11] mouse duns, and in one case a chestnut. Duns sometimes have a faint shoulder stripe, and I observed a trace of one on a bay horse. My son carefully examined and sketched for me a dun Belgian cart horse with double stripes on each shoulder as well as leg stripes. A man I can trust examined for me a small dun Welch pony with three short parallel stripes on each shoulder. Colonel Poole examined the Kattywar horse breed from the northwest part of India, and he informs me that it is so commonly striped that a horse without them is not considered purebred. The spine is always striped, legs generally barred, shoulders commonly striped (sometimes doubly and sometimes triply), and a striped face is also common. The stripes are most pronounced in the foal and sometimes disappear in old horses; Colonel Poole has observed both gray and bay Kattywar horses that were striped when first foaled. Based on information given to me by Mr. W. W. Edwards, the spinal stripe of the English racehorse is commoner in the foal than the full-grown animal. I have collected in-formation on leg and shoulder stripes in horses of different breeds from

11. "Dun" includes a large color range from one between brown and black to nearly cream.

Britain to eastern China and from Norway in the north to the Malay Archipelago in the south. In all parts of the world these stripes occur most often in duns and mouse duns.

Colonel Hamilton Smith writes that horse breeds have descended from several original species, one of which – the dun – was striped, and that the above-described appearances all result from ancient crosses with the dun stock. I disagree and resist the notion of applying his theory to breeds as distinct as the heavy Belgian cart horse, Welch ponies, cobs, the lanky Kattywar, and others.

What are the effects of crossing species within the horse genus? Rollin asserts that the common mule, from crossing an ass and a horse, is apt to have bars on its legs. I once saw a mule with legs so striped that it might have been mistaken at first for the offspring of a zebra. (In his treatise on the horse, Mr. W. C. Martin gives a figure of a similar mule.) In four color drawings I have seen of ass-zebra hybrids, the legs are more pronouncedly barred than the rest of the body, and one of them has a double shoulder stripe. Lord Moreton famously hybridized a mare chestnut and male quagga; the hybrid and even the pure offspring subsequently produced by crossing it with a black Arabian sire were more pronouncedly barred across the legs than even the pure quagga. In another remarkable case, Dr. Gray figures a hybrid between the ass and hemionus (he also knows of a second case); the ass seldom has leg stripes and the hemionus has neither leg nor shoulder stripes, and yet the hybrid has all four legs barred, three short shoulder stripes like the dun Welch pony, and even some zebra-like stripes on the side of the face. (I was so convinced that not even a stripe of color appears "by accident" that the face stripes on this hybrid led me to ask Colonel Poole whether they ever occur on the particularly striped Kattywar, and, as already stated, they do.)

So by simple variation, several distinct species of the horse genus become striped on the legs like a zebra or striped on the shoulder like an ass. This tendency is strong in the horse whenever a dun tint appears, a tint approaching the general coloring of other species in the genus. The appearance of stripes is not accompanied by any other new characteristics. The tendency to become striped is strongest in hybrids from between several very distinct species. Now recall the case of pigeon breeds; they have descended from a pigeon (including two or three sub-species or geographical varieties) of a bluish color with certain bars and other marks. When a breed assumes a bluish tint by simple variation, these bars and other marks invariably reappear, unaccompanied by any other changed characteristics. The bluish tint and bars and marks have a strong tendency to reappear in the hybrids of the oldest and truest breeds. I have stated that the most probable hypothesis to account for the reappearance of ancient characteristics is that the young of each successive generation possess a latent tendency to produce the long-lost characteristic, and this tendency sometimes prevails for unknown reasons. And, as just described, the stripes of several species in the horse genus are either more pronounced or more commonly appear in the young than in the old. If we call pigeon breeds "species" (some of them have bred true for centuries), then the case becomes exactly parallel to that of species in the horse genus! I confidently look back across thousands upon thousands of generations and see a striped animal like a zebra, perhaps otherwise differently constructed: the common ancestor of the domestic horse, whether or not it is descended from one or more wild stocks of the ass, hemionus, quagga, or zebra.

Anyone who believes that each equine species was independently created will presumably assert that each species was created with a tendency to vary the same way, both in the wild and under domestication, so as to often become striped like other species of the genus; and that each species inhabiting widely separated regions of the world was created with a strong tendency to produce hybrids resembling not the striping pattern of their own parents but that of other species in the genus. To admit this is to reject a real for an unreal, or at least an unknown. It makes the works of God a mere mockery and deception; I would as soon believe the old ignorant cosmogonists who judged that fossil shells had never lived but had been created in stone to mock the living shells of the seashore.

Our ignorance of the rules of variation is profound; not even in one case out of a hundred can we assign a reason as to why a particular part differs from the same part in the parents. But whenever a comparison *can* be made, the same rules appear to have produced the lesser differences between varieties within a species and the greater differences between species within a genus. The environment (climate, food, etc.) seems to induce some slight modifications, but habit in producing constitutional differences, use in strengthening, and disuse in diminishing organs seem to be more potent. Homologous parts tend to vary in the same way and cohere. Modifications in hard parts and external parts sometimes affect soft internal parts. A sizeably developed part might draw nourishment from adjoining parts, and every part that can be saved without detriment to the individual will be saved. Structural changes at an early age generally affect parts that develop subsequently; there are many other correlations of growth, which are fundamentally not understood. Repetitive parts are variable in structure and number, perhaps because such parts have not specialized, so their modification has gone unchecked by natural selection. This is probably the same reason that organisms low on the scale of nature are more variable than those that are high on the scale, and highly specialized. Rudimentary organs are probably variable because they are useless and therefore intractable to natural selection. Specific characteristics – those that came to differ after the species of a genus has branched off from a common ancestor – are more variable than generic characteristics – those that have not differed for a long time. In this chapter special parts or organs have been discussed that still vary because they have recently varied in order to become different. In chapter 2 the same principle was applied to the individual: on average, the most incipient species are found in regions with *many* species belonging to a particular genus, where there was much variation and differentiation previously or new species are being generated actively. Secondary sex characteristics are highly variable and differ significantly among members of the same group. Variability in the same parts generally contributes to the generation of secondary sex characteristics in the two sexes of a species and specific characteristics in species of a genus. A part or organ that is extraordinarily developed in comparison with the same part or organ in related species must have undergone an extraordinary amount of modification since the genus arose. This is why such parts or organs are still often highly variable: variation is an extended and slow process, and in these cases natural selection has not yet had time to overcome further variability or reversion. When a species with an extraordinarily developed organ becomes the common ancestor to many modified descendants – a slow process requiring a long lapse of time – natural selection fixes the organ regardless of how unusual it is. Species inheriting nearly the same constitution from a common ancestor and exposed to similar influences vary analogously and occasionally revert to some characteristics of their ancient ancestors. Although analogous variation and reversion may not generate important modifications, they add to the beautiful and harmonious diversity of nature.

Whatever the cause behind each slight difference between parent and offspring – and there must be a cause – their steady accumulation by natural selection gives rise to all the important modifications used by the innumerable beings on earth in the struggle for survival of the best adapted.

6

DIFFICULTIES WITH THE THEORY

LONG BEFORE ARRIVING AT THIS POINT IN MY WORK, A CROWD of difficulties will have occurred to the reader. Some are so grave that they stagger me to this day, but most are only apparent, and even the real ones are not fatal to my theory. These difficulties and objections can be categorized into the following groups:

1. If species have descended from other species by fine gradations, why aren't there countless transitional forms everywhere? Why isn't all nature a confusion instead of species being distinct, as they are?

2. Is it possible that an animal such as, for example, the bat, could have descended from another animal with entirely different habits? Can natural selection produce organs of slight importance, such as the giraffe's tail, which is used to shoo flies, but also produce wonderfully structured organs, such as the eye, with its barely understood and inimitable perfection?

3. Can instincts be acquired and modified through natural selection? Can the marvelous instinct of bees to make cells, practically anticipating the discoveries of mathematicians, be explained?

4. Why are the offspring of distinct species sterile while the offspring of varieties are fertile?

Topics 1 and 2 are discussed in this chapter, topic 3 is discussed in chapter 7 (Instinct), and 4 is addressed in chapter 8 (Hybridism).

Natural selection acts solely by preserving profitable modifications. In a fully stocked region, each new form tends to replace and eventually exterminate its less improved parent or other less favored forms with which it comes into competition; extinction and natural selection therefore go hand in hand. Consequently, if each species has descended from some unknown form, then its parent species and all the intervening transitional varieties will generally have become extinct by the very process of formation and perfection of the new species.

But according to this theory, innumerable transitional forms must have existed, so why aren't countless numbers found embedded in the earth's crust? I discuss this question more extensively in the chapter on the imperfection of the geological record, but the answer lies mainly in the record being incomparably less complete than generally supposed. This imperfection is due in part to the fact that organisms do not inhabit the profound depths of the sea, and in part to their remains being preserved only when embedded in sediment that is sufficiently thick and extensive to withstand an enormous amount of degradation. Such fossils only accumulate in slowly subsiding and shallow seabeds, where much sediment is deposited. These contingencies concur rarely and at enormously long intervals. The geological record leaves blanks when the seabed is stationary or rising, or when very little sediment is being deposited. Although the earth's crust is a vast museum, its natural collections were acquired at widely separated intervals of time.

Another objection is that whenever several closely related species inhabit the same region, there should be many *existing* transitional forms. Consider a simple case: in traveling north to south over a continent, closely related or representative species fill nearly the same niche at successive intervals. These representative species often meet and commingle, and as one becomes rarer, the other becomes more common. However, specimens of these species from the region where they mingle are usually as distinct in every structural detail as specimens from their respective separate habitats. By my theory, related species descend from a common parent; during the process of modification, each becomes adapted to the environment of its own region and supplants its parent and the transitional varieties between its past and present states. Therefore, we should not expect to find numerous transitional varieties living in each region, even though they must have once existed and may be embedded there as fossils. The next question is, why are intermediate varieties absent in the intermediate region, with its intermediate environment? This question long confounded me, but I think it can in large part be explained.

In the first place, just because a region is continuous now does not mean it has been continuous over a long period. Geology suggests that almost every continent has been broken up into islands, even during the later Tertiary periods. On such islands, distinct species might have formed without intermediate varieties, intermediate zones having been absent. Marine areas that are now continuous often must have existed less continuously and uniformly within recent times due to changes in climate and the form of the land; but I will pass over this way of escaping the difficulty, because many perfectly defined species have formed on continuous areas. (I believe, however, that the formerly fragmented condition of areas that are now continuous played an important role in the formation of new species, especially with freely crossing and wandering animals.)

Widely distributed species are generally numerous over a large territory and then abruptly thin out near the edges. Therefore the overlapping territory between two representative species is generally narrower than their proper territories. The same phenomenon is observed in the ascent of mountains; as Alph. de Candolle observes, common alpine species sometimes disappear remarkably abruptly above a particular range. Forbes also notices the same thing in sounding the depths of the sea with a dredge. These observations should surprise anyone who looks to climate and the physical environment as defining distribution, because these factors change gradually rather than abruptly. But recall that almost every species would proliferate immensely, even in its proper territory, were it not for competing species, and that all species either prey on or serve as prey to others: each organism is directly or indirectly related to other organisms. So the range of a species is not defined exclusively by graded physical conditions but in large part by other species on which it depends, by which it is destroyed, or with which it competes. These other species are already defined objects (regardless of how they have become defined) that do not blend into one another gradually, *so a species' range is sharply defined because it depends on the ranges of others.* Furthermore, at the edge of its range each species exists in lower numbers, so during fluctuations in the number of its enemies, prey, or in the seasons, individuals in this border region are very susceptible to complete extermination, thus defining geographical range even more sharply.

Because varieties do not essentially differ from species, the above concepts concerning distribution probably apply to varieties as well. Imagine two varieties adapted to two large areas and a third variety to a narrow intermediate zone. The intermediate variety will be rarer from inhabiting a narrow and smaller area; as far as I can make out these rules are true in the wild. I have found striking examples of the rule in cases of varieties that are intermediate between well-defined varieties in the genus *Balanus.*[1] Information provided by Mr. Watson, Dr. Asa Gray, and Mr. Wollaston suggests that intermediate varieties are generally fewer in number than the forms they connect. If these observations are correct, they illuminate why intermediate varieties do not endure for long periods and usually become exterminated sooner than the forms they originally linked.

A rare form runs a greater chance of extermination than a common form, and an intermediate form is liable to suffer the inroads of closely related forms from both sides. More important, during the process of modification by which varieties become distinct species, those existing in large numbers over large areas have a huge advantage over the intermediate variety, which exists in low numbers in a narrow and intermediate zone. A large population is always more likely than a small population to present favorable variations for natural selection to seize. In the race for life, the more common forms tend to beat and supplant the less common forms, because these are modified and improved more slowly. As discussed in chapter 2, this same principle accounts for common species of a region having more varieties than the rarer species. To illustrate, imagine three varieties of sheep: one adapted to an extensive mountainous region; a second to a comparatively narrow, hilly tract; and a third to wide plains. The human inhabitants all try with equal skill and steadiness to improve their stocks by selection. In this case, those in the

---

1. [*Balanus* is a genus of barnacle. – D.D.]

mountains and the plains are much more likely to improve their breeds than those in the narrow, hilly tract; the improved mountain or plain breed will consequently replace the less improved hill breed, and the two breeds that were originally more numerous will come into close contact without the interposition of the supplanted hill variety.

In summary, I believe that species become fairly well defined objects and do not form an inextricable chaos of varying and intermediate links because:

1. Varieties form very slowly (variation is a slow process). Natural selection can do nothing until favorable variations happen to occur and until a niche can be better filled by one or more modified varieties. The formation of such new places depends on slow climatic changes, on occasional immigration of new inhabitants, and probably on some old inhabitants becoming slowly modified with new and old forms acting and reacting on each other. Therefore a particular region can maintain only a few species with slight modifications of structure that are in some degree permanent – and this is indeed observed.

2. Areas that are now continuous have often existed within the recent period as isolated regions on which many forms may have been rendered sufficiently distinct to rank as representative species; this is especially likely among organisms that mate to reproduce and wander extensively. In this case, varieties intermediate between the representative species previously existed on each fragment of land, but these links have been supplanted during the process of natural selection and no longer exist.

3. When two or more varieties develop in different parts of a continuous area, intermediate varieties probably start to develop in the intermediate zone, but they generally do not last long. Based on what we know about the actual distribution of closely related or representative species and acknowledged varieties, these intermediate varieties will be rarer than those they connect. This alone predisposes the intermediate varieties to accidental extermination, and during the process of further

modification through natural selection they will almost
certainly be replaced by the forms they connect, which have
larger populations and therefore harbor more variations for
natural selection to act upon and generate advantages.

4. If my theory is correct, then throughout the history of life on
   earth, countless intermediate varieties have existed, closely
   linking related species. However, the very process of natural
   selection tends to exterminate parental forms and intermediate
   links, so only fossilized evidence of their former existence
   could possibly be found. However, as I will discuss in a future
   chapter, fossil remains are preserved in an extremely imperfect
   and intermittent record.

Opponents of the views I hold ask, how could a carnivorous land animal,
for example, convert into an aquatic animal? After all, how could the
animal survive in its transitional state? But it would be easy to show that
within a single group there are carnivorous animals with every characteristic between truly aquatic and strictly terrestrial. Each exists through
a struggle for life and is well adapted to its niche. Consider the *Mustela
vison* of North America with its webbed feet, otter-like fur, short legs, and
tail.[2] It dives and preys on fish during the summer, but during the long
winter it leaves the frozen waters and, like other polecats, preys on mice
and land animals. Had the question been, "How could an insectivorous
quadruped convert into a flying bat?" it would have been far more difficult, and I could have given no answer. But as I will now show, I think
such difficulties have very little weight.[3]

Consider the squirrel family, with its fine gradations from animals
with slightly flattened tails, through those with relatively wide posterior
sections and full skin on their flanks (as Sir J. Richardson has remarked),

2. [The *Mustela vison* is a species of mink, belonging to the same family as weasels,
otters, etc. – all of them carnivores. – D.D.]

3. As on other occasions I am at a disadvantage, because out of the many striking
cases I have collected, only one or two examples can be given of transitional structures
and habits in closely related species of the same genus and of diversified habits (constant
or occasional) of the same species. Nothing short of a long list is sufficient to lessen the
difficulty in any particular case, like that of the bat.

to flying squirrels with their limbs and even the bases of their tails united by a broad expanse of skin. (It serves as a parachute and enables the flying squirrel to glide astonishing distances through the air from tree to tree.) Each structure is useful to each kind of squirrel in its own region by enabling it to escape from predators, collect food more quickly, or, apparently, by reducing the danger from occasional falls. *But it does not follow that the structure of each squirrel is the best possible under all environmental conditions.* If the climate or vegetation should change; if other, competing rodents or new predators should immigrate; or if old competitors or predators should become modified, then all analogy suggests that some squirrels would decrease in number or become exterminated unless they also became modified and improved in structure. Therefore I can see no difficulty in the idea that individuals with fuller and fuller flank membranes could be preserved, especially in a changing environment. Each modification would be useful, so each modification would be propagated until the cumulative effect of natural selection produced a perfect flying squirrel.

Now consider the flying lemur, once incorrectly considered a bat. It has a very wide flank membrane, stretching from the corners of the jaw to the tail and including the limbs and elongated fingers; the membrane is even furnished with an extensor muscle. Although no graduated links now connect the flying lemur to other lemur species, there is no difficulty in supposing that they once existed and that each was formed by the same steps as applied in the case of the less perfectly gliding squirrels; each grade of structure was useful to its possessor. There is no difficulty in further supposing that the membrane-connected fingers and forearm of the flying lemur could be lengthened by natural selection, and, as far as flight organs are concerned, this would render it a bat. Indeed, in bats having wing membranes that extend from the top of the shoulder to the tail, including the hind legs, we may perhaps discern traces of an apparatus originally constructed for gliding rather than flying.

If about half a dozen genera of existing birds had become extinct or were unknown, who would have guessed that birds might use wings solely as flappers (like the logger-headed duck), as fins in the water but front legs on land (like the penguin), as sails (like the ostrich), and for no functional purpose (like the kiwi)? Yet the wing structure of each of these birds is useful in the particular environment in which it lives, for each lives by a struggle. However, a given wing structure is not neces-sarily best under all possible conditions. These examples, which may all have resulted from disuse, do not illustrate the natural steps by which birds acquired their perfect ability to fly, but rather how diverse the means of transition can be.

Because several members of water-dwelling classes like the crusta-ceans and mollusks are adapted to live on land, and because there are fly-ing birds, mammals, insects, and (now extinct) reptiles, it is conceivable that flying fish, which glide far through the air, rising and turning with the aid of fluttering fins, might become modified into perfectly winged animals. If this happened, who would guess that in an early transitional state they had inhabited the open ocean and used their incipient organs of flight to escape being devoured by other fish?

Bear in mind that if an organism already possesses a structure that is highly perfected for a particular activity (like the wings of birds for flight), then animals with early transitional versions of the structure will seldom still exist, because they will have been supplanted by the very process of perfection through natural selection. Furthermore, the chance of discovering fossils with transitional structures is low, because species with such structures would have existed in smaller numbers than species with fully developed structures.

When the habits of a species change or diversify, natural selection can easily fit the species to its changed habits or exclusively to one of its several different habits by some modification of its structure. It is dif-ficult to tell (and irrelevant to the argument) whether habit generally changes first and structure afterward or vice versa; in most cases, both probably change almost simultaneously. A sufficient example of changed habits is the set of many British insects that now feed on exotic plants or exclusively on artificial substances. And there are countless examples of diversified habits. I have often observed the tyrant flycatcher in South America hovering over one spot and then proceeding to another, like a kestrel, and at other times standing stationary on the margin of the water before dashing at a fish like a kingfisher. In Britain the larger titmouse sometimes climbs branches almost like a creeper, kills small birds by blows to the head like a shrike, and I have many times seen and heard it hammering the seeds of the yew on a branch and thus breaking them like a nuthatch. In North America, Hearne has observed the black bear swim for hours with an open mouth, like a whale, to catch insects in the water. If the insect supply were constant, and if competitors better adapted than the bear did not already exist in the region, then a variety of bear could be rendered more and more aquatic and larger and larger mouthed by natural selection until a creature was produced as monstrous as a whale.

Individuals sometimes have very different habits from other members of their species or genus. My theory predicts and nature shows that such individuals do occasionally give rise to new species with slightly or considerably modified structures. The woodpecker provides a striking example. It is perfectly adapted for climbing trees and seizing insects in the chinks of bark, yet there are woodpeckers in North America that feed mostly on fruit and others with elongated wings that chase insects on the wing. On the plains of La Plata, where not a tree grows, there is a woodpecker that is like the common woodpecker in its coloring, in every essential part of its organization, even in its harsh tone and undulatory flight – yet it never climbs trees!

Petrels are aerial oceanic birds, yet in the quiet sounds of Tierra del Fuego, the *Puffinuria berardi* resembles an auk or grebe in its general habits, astonishing diving abilities, manner of swimming, and flight when it unwillingly takes to the air. It would be easy to mistake it for an auk, but it is nevertheless a petrel with many parts of its organization profoundly modified. Conversely, even the most careful observer of a dead water ouzel's body would never suspect its aquatic habits, yet this anomalous member of the terrestrial thrush family subsists entirely by diving – grasping stones with its feet and using its wings underwater.

Anyone who believes that each organism has been created as we now see it must be surprised by animals with structures that are in disagreement with their habits. The webbed feet of ducks and geese are obviously for swimming, yet upland geese with webbed feet rarely or never go near water. And no one except Audubon has seen the frigate bird alight on the surface of the sea, yet all its four toes are webbed. Conversely, grebes and coots are aquatic, but their toes are bordered only by a membrane. The long toes of grallatores are clearly formed for walking over swamps and floating plants, yet the water hen is almost as aquatic as the coot, and the land rail almost as terrestrial as the quail or partridge. In these and many other cases, habits have changed without a corresponding change in structure. The webbed feet of the upland goose have become rudimentary in function, though not in structure, while the deep scoop of the membrane between the toes of the frigate bird shows that its structure has begun to change.

Anyone who believes in countless separate acts of creation would claim that in these cases it has pleased the Creator to cause one type of organism to take the place of another, but this is simply a restatement of the observation in dignified language. Anyone who believes in the struggle for existence and in the principle of natural selection will acknowledge that all organisms constantly endeavor to increase their numbers. Thus, any individual with even a slight advantageous variation in habit or structure over another inhabitant of the region will seize that inhabitant's place, however different it may be from its own. So it is unsurprising that there should be geese and frigate birds with webbed feet, either on dry land or rarely alighting on water; that there should be long-toed corncrakes living in meadows instead of in swamps; that there should be woodpeckers where not a tree grows; that there should be diving thrushes, and petrels with the habits of auks.

To suppose that the eye – with its inimitable contrivances for focusing objects at different distances, admitting different amounts of light, and correcting for spherical and chromatic aberration – could have been formed by natural selection seems absurd. But reason tells me that actually – though it seems so hard to imagine – the difficulty is not real. Natural selection can indeed act as the mechanism for the formation of a perfect and complex eye if the following three conditions are met: (1) if we can show that there are numerous gradations from an imperfect and simple eye to a perfect and complex eye, and that each intermediate form is useful to its possessor; (2) if the eye does vary, even a little, and those variations can be inherited, which is certainly the case; and (3) if any variation or modification in the organ is ever useful to an animal in a changing environment. How a nerve becomes sensitive to light does not concern us any more than how life originated, but several observations suggest that any sensitive nerve can be rendered responsive to light – or to the coarser vibrations of the air, which are responsible for sound.

In looking for the gradations by which any organ in any species has been perfected, we ought to look exclusively to direct ancestors. This is rarely possible, so instead we are forced to consider species from the same group (collateral descendants from the same common ancestor) to see what gradations are possible. This requires some gradations having been transmitted unaltered, or only slightly altered, from early stages of descent. Among existing vertebrates there is a small amount of gradation in the structure of the eye, and fossils do not provide any information about this organ. In this great class we would probably have to descend far beneath the lowest known fossil deposits to discover the early stages by which the eye has been perfected.

We can begin a series with the arthropods.[4] Some of them have an optic nerve coated with pigment and no other mechanism. From this low stage, numerous gradations of structure branch off in two fundamentally different lines until a moderately high level of perfection is reached. For example, certain crustaceans have a double cornea. The inner one is divided into facets, within each of which there is a lens-shaped swelling. In other crustaceans the transparent pigment-coated cones, which act only by excluding lateral pencils of light, are convex at their upper ends and must function by converging light; there seems to be an imperfect vitreous substance at their lower ends. These brief observations demonstrate that there is much graduated diversity in the eyes of living crustaceans. Bearing in mind the small number of *living* organisms relative to the huge number of *extinct* organisms, there is no great difficulty (no greater than for any other structure) in believing that natural selection has converted the simple apparatus of an optic nerve coated with pigment and invested with a transparent membrane into an optical instrument as perfect as any possessed by an arthropod.

Anyone who finishes reading this book and finds that large sets of otherwise inexplicable facts can be explained by the theory of descent should not hesitate to go further and admit that even a structure as perfect as the eye of the eagle might be formed by natural selection. (I have felt the difficulty of extending natural selection to such startling lengths far too keenly to be surprised by any amount of hesitation.) In this case the transitional grades are unknown, but reason should conquer imagination.

4. [Arthropods are broadly defined as invertebrates with segmented limbs, including, for example, insects and crustaceans. – D.D.]

It is difficult to avoid comparing the eye to the telescope. We know that this instrument has been perfected by the long and continuous efforts of the highest human intellects, and we naturally infer that the eye has been formed by an analogous process. But isn't this inference presumptuous? Have we any right to assume that the Creator works by intellectual powers like those of man? If we *must* compare the eye to an optical instrument, we should imagine a thick layer of transparent tissue with a light-sensitive nerve beneath. We would then have to imagine that every part of this layer constantly and slowly changes in density, so there are separate layers of differing density, thickness, and distance from one another; the surface of each layer slowly changes form. Some power would always have to intently watch every slight accidental alteration in the transparent layers and select the alterations that under various circumstances give more distinct images. Imagine each new state of the instrument multiplied by millions and preserved until a better one is produced and the old ones destroyed. In living organisms, variation causes the slight alterations, generations multiply them almost infinitely, and natural selection unerringly picks out each improvement. Let this process go on for millions and millions of years and during each year on millions of individuals of many kinds, and may we not believe that this can result in a living optical instrument that is superior to one of glass, as the works of the Creator are to those of man?

If there is a complex organ that could not possibly have been formed by numerous, successive, slight modifications, then my theory absolutely breaks down. But I can find no such case. No doubt there are many organs for which the transitional grades are unknown, especially among very isolated species around which, according to my theory, there has been much extinction. And an organ common to all members of a large class must have first formed a very long time ago; discovering its early transitional grades would require looking at very ancient ancestral forms, long since extinct.

We should be extremely cautious in concluding that an organ could not have been formed by gradations. In several lower animals, one organ performs multiple distinct functions. For example, the alimentary canal of the dragonfly larva and the fish *Cobites* digests, respires, *and* excretes. And if a hydra is turned inside out, then what was at first the outer surface will digest and what was the stomach will respire! Natural selection can easily specialize a part or organ with two functions for one function alone if there is any advantage to the change, thus by small steps entirely altering its nature.

Sometimes two distinct organs perform the same function simultaneously. For example, there are fish with gills that breathe the air dissolved in water and at the same time breathe free air in their swim bladder, which is supplied by a pneumatic duct and divided into highly vascular portions. In such cases one of the two organs can easily be modified and perfected to perform all the work by itself, aided during modification by the other organ, which might eventually be modified for some other unrelated purpose – or even obliterated.

The example of the swim bladder clearly illustrates that an organ originally constructed for one purpose, such as flotation, can be converted into an organ for an entirely different purpose, such as respiration. In certain fish the swim bladder has also been worked in as an accessory to the auditory organs, or possibly vice versa. Physiologists agree that the swim bladder is homologous ("ideally similar") in position and structure with the lungs of higher vertebrates. So there is no difficulty in believing that natural selection has converted a swim bladder into a lung, an organ used exclusively for respiration. In fact, I am convinced that all vertebrates with true lungs have descended from an unknown ancient prototype – of which nothing is known – with a floating apparatus or swim bladder. This explains the strange fact that every particle of food and drop of drink has to pass over the tracheal orifice with some risk of falling into the lungs, despite the beautiful contrivance by which the glottis closes. Gills have disappeared from higher vertebrates, yet the slits on the sides of the neck and the looped course of the arteries in the embryo still mark their former position. It is conceivable that the now-lost gills have been gradually worked in by natural selection for some new purpose. Some naturalists maintain that the gills and dorsal scales of annelids are homologous with the wings and wing covers of insects;[5] thus, organs that at a very ancient period served for respiration have probably been converted into organs of flight.

5. [Annelids are segmented worms, such as earthworms. – D.D.]

The matter of how organs become transformed from one function to another is so important that I will give one more example. Stalked barnacles have two minute folds of skin, the *ovigerous frena,* which retain the eggs in a sticky secretion within the sac until they hatch. These barnacles have no gills; instead, the whole surface of the body and the sac respire, including the small *frena.* In contrast, rock barnacles have no *ovigerous frena,* so the eggs lie loosely at the bottom of the sac in the enclosed shell, but they have large folded gills. The *ovigerous frena* of the one family are obviously homologous with the gills of the other; indeed, they graduate into one another. Little folds of skin that originally served as *ovigerous frena* but also contributed to respiration have gradually been converted by natural selection into gills simply through an increase in their size and the elimination of their adhesive glands. If all stalked barnacles had become extinct (and they have already suffered far more extinction than rock barnacles), who would ever have imagined that the gills of the rock barnacle family were originally organs used to prevent eggs from being washed out of the sac?

A particularly challenging case is that of neuter insects, which are often constructed differently from males and fertile females, but this will be discussed in the next chapter. Electric organs of fishes present another difficult case, as it is impossible to conceive by what steps these wondrous organs have arisen. But as Professor Owen and others have remarked, their structure closely resembles that of common muscle. Moreover, rays have an organ that is similar to the electric apparatus, and yet, as stated by Matteuchi, they do not discharge electricity.

Another, even more serious difficulty is that only about a dozen fish species have electric organs, and several of these are very distantly related. If the same organ appears in several members of the same class, especially if these members have very different habits, then its presence can generally be attributed to inheritance from a common ancestor, and its absence in some of the members to loss through disuse or natural selection. Had the electric organs been inherited from a single common ancestor, then all electric fishes should be specially related. Fossils do not suggest that most fishes once had electric organs that most of their descendants subsequently lost. The luminous organs found in some insects belonging to different families and orders offer a parallel case of difficulty. Similarly in plants, the curious structure of pollen borne on a stalk with a sticky gland at the end is the same in *Orchis* and *Asclepias*, genera that are almost as remote as possible among flowering plants. An important consideration applies to all such cases in which very distinct species possess apparently the same anomalous organ: although the overall appearance and function of the organ may be the same, there is generally some fundamental difference. Just as two people sometimes independently hit upon the same invention, natural selection sometimes modifies two parts in two very distantly related organisms nearly the same way. It works for the good of each and takes advantage of analogous variations. We are far too ignorant to argue that no transition of any kind is possible.

Although in many cases it is difficult to conjecture by what transitions an organ could have arrived at its present state, surprisingly few can be named toward which *no* transitional grade is known. (This is even more surprising considering that the number of living forms is far less than the number of extinct forms.) The absence of transitional grades is reflected in that old maxim of natural history, *Natura non facit saltum*,[6] encountered in the writings of almost every experienced naturalist. Milne Edwards expressed it by stating that nature is prodigal in variety but niggardly in innovation. Based on the theory of creation, why should this be the case? Why should all the supposedly separately and purposefully created parts and organs of many independent organisms be linked by graduated steps? Why didn't nature leap from structure to structure? The theory of natural selection clearly explains why not: natural selection can act only by taking advantage of slight successive variations. Advances cannot be made by leaps, but by the shortest and slowest steps.

Natural selection acts through life and death, by the preservation of individuals possessing any favorable variation and the destruction of individuals with unfavorable deviations of structure. It is therefore difficult to understand the origin of simple parts that do not seem important enough for preservation in successive varying individuals. I have

6. ["Nature does not leap." – D.D.]

sometimes felt as much difficulty – though of a different kind – as in the case of an organ that is as perfect and complex as the eye.

Far too little is known of the entire machinery of any one organism to decide which slight modifications are important or unimportant. In chapter 4 I gave examples of minor characteristics, such as the fuzz and flesh of fruit, that are correlated with constitutional differences or determine whether insects attack and can thus be acted on by natural selection. It seems incredible that the giraffe's tail, which looks like an artificially constructed fly flapper, could have been adapted for its present purpose by small successive modifications, but caution is necessary even in this case, because the distribution of cattle and other animals in South America depends entirely on their abilities to resist insect attacks. Individuals capable of defending themselves from these small enemies can range into new pastures – a huge advantage. Not that large animals are necessarily destroyed by flies (except in rare cases), but they are incessantly harassed and their strength sapped. They become more subject to disease and less able to search for food in an impending famine or to escape from predators.

Organs that are now of little importance may have been significant in an early ancestor, slowly perfected in the past and transmitted in the same state. Any harmful deviations of such an organ would have been checked by natural selection. In aquatic animals the tail is an important organ of locomotion. So perhaps it is easy to understand why land animals – which are descended from aquatic animals, as their gills or modified swim bladders make clear – usually have tails and also why this organ (no longer useful for locomotion) has become modified for many other uses. A tail originally formed in an aquatic animal might subsequently have been developed for all sorts of purposes, like a fly flapper, an organ of prehension, or an aid in turning, as in dogs (although its use for this must be minimal, because the hare has hardly any tail at all and it can double quickly enough).

We may sometimes give too much importance to characteristics that are really rather trivial and that have originated from secondary causes independently of natural selection. Recall that climate, food, and other external factors probably have some minor direct effect on organization; that characteristics reappear from reversion; that correlated growth influences the modification of various structures; and that sexual selection often extensively modifies the external characteristics of males for fighting one another or for charming females. Furthermore, when a modification in structure arises from these or other unknown causes, it may not provide an initial advantage, but it may eventually become advantageous in descendants living in new environments with new habits.

To illustrate, imagine if only green woodpeckers existed and black and pied kinds were unknown. We would think that their green color is a beautiful adaptation to hide them in trees from enemies and therefore important and acquired through natural selection. In reality, however, the color is due to some other cause, probably sexual selection. The trailing bamboo of the Malay Archipelago climbs the loftiest trees with the aid of exquisitely constructed hooks clustered around the ends of branches; these are obviously useful to the plant, but many trees that are not climbers have similar hooks, so those on the bamboo may have arisen from unknown rules of correlated growth and then been taken advantage of as the plant became modified into a climber. The naked skin on the head of a vulture is generally considered an adaptation for wallowing in rotten meat, or it might result directly from the effects of the putrid matter, but any such inference should be drawn cautiously, given that the skin on the head of the male turkey is also naked even though this bird feeds cleanly. The sutures in the skulls of young mammals have been advanced as a beautiful adaptation for aiding birth, and no doubt they facilitate and may even be indispensable for it. But sutures also occur in the skulls of young birds and reptiles, which have only to escape from eggs, so we infer that sutures arose from correlated growth and have been taken advantage of to aid birth in higher animals.

The causes producing slight and unimportant variations are unknown. Consider the differences between domestic animal breeds from different regions, especially where there has been limited artificial selection. Careful observers are convinced that a damp climate affects the growth of hair, which is correlated with the growth of horns. Mountain breeds always differ from lowland breeds; a mountainous region probably affects the hind limbs by exercising them more and maybe even affects the shape of the pelvis. And then by homologous variation, the front limbs and even the head would probably be affected. By the way it exerts pressure, the shape of the pelvis may also affect the shape of the head in the womb. The laborious breathing necessary at high altitudes would probably increase chest size, and correlation would again come into play. Animals kept by natives in different regions often struggle for their own food and would be exposed to some natural selection; individuals with slightly different constitutions would succeed best under different climates, and constitution and color are correlated. In cattle, both the susceptibility to attacks by flies and the liability to be poisoned by certain plants are correlated with color, so color is subjected to natural selection. Too little is known to speculate on the relative importance of the various known and unknown rules of variation. By alluding to them here, I mean to show only that if we cannot account for the characteristic differences among domestic breeds (though they arise by ordinary generation), we should not overstress our ignorance of the precise causes of slight analogous differences among species. I could also have mentioned the differences among human races, which are so strongly marked; the origins of these differences can apparently be somewhat illuminated by sexual selection of a particular kind, but without entering into extensive details my reasoning would appear frivolous.

These remarks lead me to briefly discuss the recent protest of some naturalists against the utilitarian doctrine that every detail of structure exists for the good of its owner. They believe that many structures have been created for the beauty of human eyes or simply for variety; however, if this were true, it would be fatal to my theory. Yet I admit that many structures are of no direct use to their owners. The physical environment probably has some effect on structure independently of any advantage gained. Correlated growth plays an important role such that a useful modification of one part often entails changes of no direct use in other parts. Formerly useful characteristics, or characteristics that arose from correlated growth or other unknown causes, may reappear through reversion, though they are no longer of any direct use. The effects that sexual selection has on beauty – so as to charm the females – are not useful in the strictest sense. However, by far the most important consideration is that the main organization of every living thing is due to inheritance. Consequently, though every organism is fit to its current niche, many have structures that are no longer directly related to their habits. The webbed feet of the upland goose and frigate bird are of no special use to them. Likewise, the same bones in the arm of the monkey, foreleg of the horse, wing of the bat, and flipper of the seal are of no special use to these animals: the structures result from inheritance. But webbed feet were definitely as useful to the ancestor of the upland goose and frigate bird as they are now to most aquatic birds. Similarly, the ancestor of the seal did not have a flipper, but rather a foot with five toes fit for walking or grasping. The limb bones of the monkey, horse, and bat have been inherited from a common ancestor or ancestors that had a more specialized use for them than these animals with such diversified habits. These bones were acquired through natural selection, subjected then and now to inheritance, reversion, correlated growth, and so on. And making some little allowance for direct environmental action, every detail in the structure of every living creature either had a special function in some ancestral form or now has a special function in its descendants, either directly or indirectly through the complexity of correlated growth.

Natural selection cannot possibly produce any modification in one species exclusively for the good of another species, although throughout nature, species take advantage of and profit by the structures of other species. But natural selection can and does produce structures that directly harm other species, such as the fang of the adder and the ovipositor of the ichneumon, by which eggs are deposited in the living bodies of other insects. My theory would be annihilated if any structure of any species were proven to have been formed for the exclusive good of another species; such a part cannot form through natural selection. Although many claims to this effect are found in works of natural history, I cannot find even one with any weight. The rattlesnake has a poison fang for defense and for the destruction of prey, but some authors also suppose that the snake has a rattle to its own disadvantage – namely, to warn its prey to escape. I would as soon believe that when preparing to pounce, the cat curls the end of its tail to warn the doomed mouse! There is not enough space here to discuss these and other such cases.

Natural selection acts only by and for the good of each organism and never produces an organ harmful to its owner. As Paley has remarked, no organ is formed for the purpose of inflicting pain or harm on its owner. Each part is advantageous overall when its good and bad effects are fairly weighted. If with the passage of time a changing environment renders a part harmful, it will be modified; if not, the organism will become extinct, as so many have.

Natural selection tends to make each organism at least as well adapted as its competitors; this is the degree of perfection attained in nature. For example, the endemic organisms of New Zealand are perfectly adapted in one another's company but are rapidly yielding before the advancing legion of plants and animals introduced from Europe. Natural selection does not produce absolute perfection, and as far as can be judged this standard does not even exist. The correction for the aberration of light is imperfect even in that most perfect organ, the eye. If our reason leads us to admire with enthusiasm many inimitable structures in nature, then it also leads us to recognize that many others are imperfect (although we might easily err on both sides).

The backward serrations of the wasp or bee's stinger prevent its withdrawal from a victim, inevitably killing its wielder by tearing out the viscera. Is it "perfect"? Yet this extreme imperfection makes sense when we consider that in a remote ancestor the bee's stinger may originally have been a serrated boring instrument – something that many existing members of the same order still have. The stinger has been modified but not perfected for its present purpose; the poison that was originally adapted to cause galls in plants has been intensified. This may explain why stinging often causes the insect's own death: if stinging ability is generally useful to the whole community, it fulfills the requirements of natural selection even though it may cause some members to perish. If we admire the wonderful ability of the males of many insect species to find females by scent, we should admire the thousands of drones that are utterly useless to the community for anything else and are ultimately slaughtered by their industrious and sterile sisters. Although it may be difficult, we should admire the queen bee's savage instinct to destroy her queen daughters or perish herself in the combat, because this is best for the community. Maternal love and maternal hatred – though the second is fortunately rare – are all the same to the relentless principle of natural selection. If we admire the ingenious mechanisms that facilitate pollination by insects in the flowers of orchids and many other plants, then we can admire as equally perfect the way that fir trees release dense clouds of pollen so that a few granules might be wafted by a chance breeze on to the ovules.

This chapter has covered some of the difficulties with and objections to my theory. Many of them are serious. At the same time, I have presented several observations that are completely inexplicable under the theory of independent acts of creation. Species are not indefinitely variable and not linked together by many intermediates at any one given time. This is partly because natural selection is very slow and acts on only a few forms at a time, and partly because natural selection implies the continued replacement of preceding and intermediate forms. Closely related species now occupying a continuous area were often formed when the area was not continuous and when the environment did not graduate from one part to another. When two varieties develop in two regions of a continuous area, an intermediate variety often develops in an intermediate zone. The intermediate variety is usually rarer than the forms it connects, which become further modified because of the advantage of their greater numbers; they thus tend to supplant and exterminate the intermediate variety.

This chapter showed that even the most different habits of life can graduate into one another. For example, a bat can develop through natural selection from an animal that at first could only glide through the air.

In a new environment a species can change or diversify its habits, with some habits very different from those of its nearest relatives. This explains how there are upland geese with webbed feet, ground woodpeckers, diving thrushes, and petrels with the habits of auks.

Believing that an organ as perfect as the eye has formed through natural selection is enough to stagger anyone. However, if a long series of useful gradations of any organ is known, then there is no logical impossibility in any amount of perfection through natural selection. Just because in some cases intermediate or transitional states are unknown does not mean they never existed; the homologies of many organs and their intermediate states show that wonderful functional metamorphoses are possible. For example, the swim bladder has apparently been converted into an air-breathing lung. An organ can simultaneously perform very different functions and then become specialized for one function, or two organs can perform the same function and then one can become perfected while being aided by the other. Both of these mechanisms must have often greatly facilitated transitions.

In most cases we are far too ignorant to assert that a part or organ is so unimportant that modifications of its structure could not have been slowly accumulated by natural selection. But many modifications that have resulted entirely from correlated growth and were at first of no advantage to a species have subsequently been seized by still further modified descendants. A once very important part can be retained even if it could not have been acquired in its present, less important state through natural selection, which acts solely by preserving variations that are profitable in the struggle for life.

Natural selection produces nothing in one species for the exclusive good or injury of another species. It can produce parts or organs that are very useful or even indispensable, or very harmful to another species, but only if they are useful to the owner. If a region is densely inhabited, then natural selection acts chiefly through the competition of inhabitants with one another; it consequently produces perfection – strength in the struggle for existence – according to the standards of that region. This is why inhabitants of a small region generally yield to inhabitants of a larger region, where there are more individuals, greater diversification, more severe competition – and as a result, a higher standard of perfection. Natural selection does not necessarily produce absolute perfection, which, as far as can be judged, does not even exist.

The theory of natural selection clearly illuminates that old maxim of natural philosophy, *Natura non facit saltum*. This saying is not strictly correct if only the earth's present inhabitants are considered, but if all those of past times are included, it must, by my theory, be strictly true.

"Unity of type" and "conditions of existence" are generally acknowledged as governing the formation of all organisms. Unity of type is the fundamental similarity in the structures of organisms within a class, regardless of their habits. In my theory, unity of type is explained by unity of descent. The concept of "conditions of existence," so often asserted by the illustrious Frederick Cuvier, is fully embraced by the principle of natural selection; natural selection either adapts the varying parts of each organism to its environment or has already adapted them during long-past periods of time. The adaptations are sometimes aided by use and disuse, are slightly affected by direct environmental action, and are always subjected to correlated growth. Therefore, the concept of "conditions of existence" *includes,* through the inheritance of former adaptations, "unity of type."

7

INSTINCT

INSTINCT COULD HAVE BEEN WORKED INTO PREVIOUS CHAP-
ters, but I thought it would be better to treat it separately, especially because an instinct as wonderful as that of the honeybee making the cells of
its hive may have occurred to many readers as sufficiently challenging to
overthrow my whole theory. I must premise that my theory has nothing
to do with the origin of primary mental powers, just as it has nothing to
do with the origin of life itself; this discussion is concerned only with the
diversity of instincts and other mental qualities of animals within a class.

I will not try to define instinct. Several distinct mental capacities are
commonly embraced by the term, but everyone knows what is meant by
instinct impelling a cuckoo to migrate and lay her eggs in other birds'
nests. An action that in humans would require experience, when performed by animals (especially young ones) and by many individuals
in the same way without their knowing its purpose is usually said to be
instinctive. But none of these characteristics of instinct is universal. As
Pierre Huber expresses it, a small dose of reason or judgment is often
involved even with animals low on the scale of nature.

Frederick Cuvier and several of the older metaphysicians compare
instinct with habit, accurately highlighting the mental state under which
instinctive actions are performed, but not their origin. Many habitual
actions are performed unconsciously – sometimes in direct opposition
to conscious will – yet they can be modified by will or reason. Habits
easily become associated with other habits, periods of time, and bodily
states; they often remain constant throughout life. Several other similarities between instincts and habits could be pointed out. One action

of an instinctive behavior follows another by a sort of rhythm, like when a person recites a well-known song; if he is interrupted while singing or while repeating anything by rote, he goes back to recover the habitual train of thought. Huber studied a caterpillar that makes a complex hammock. When he took a caterpillar that had completed its hammock up to the sixth stage of construction and put it into a hammock completed only to the third stage, the caterpillar re-performed the fourth, fifth, and sixth stages of construction. When a caterpillar was taken out of a hammock made up to the third stage and put into one finished to the sixth stage, instead of enjoying the benefit it started from the third stage, where it had left off, in an attempt to complete the already finished work.

If a habitual action becomes heritable, which sometimes does happen, then the resemblance between what was originally a habit and what was instinct becomes indistinguishably close. If Mozart had played the pianoforte with no practice at all, instead of at age three with exceptionally little practice, then he might truly be said to have played instinctively. But it would be incorrect to assume that most instincts are acquired by habit in one generation and then transmitted by inheritance to succeeding generations. The most wonderful known instincts are those of the honeybee and many ants, and they could not possibly have been acquired this way.

Instincts are as important to the welfare of each species as bodily structure. If environmental conditions change, then slight modifications of instinct might be profitable to a species. If instincts vary even slightly, then natural selection can preserve and accumulate these variations to any extent that is profitable. This is how all the most complex and wonderful instincts have originated. Just as modifications of bodily structure arise and develop by use or habit and diminish or disappear by disuse, so do modifications of instinct. However, the effects of habit are subordinate to the effects of the natural selection of "accidental variations of instinct" (i.e., variations of instinct produced by the same unknown causes that generate variations of bodily structure).

Natural selection can produce complex instincts only by the slow and gradual accumulation of slight yet profitable variations. Just as with bodily structures, we cannot expect to find the actual transitional gradations by which each complex instinct has been acquired, because these existed only in the direct ancestors of each species. Instead, we should find some evidence of gradations in collateral lines of descent, or at least discern that such gradations are possible (this can definitely be done). I am surprised by how generally gradations leading to complex instincts can be discovered, even allowing for the instincts of animals barely observed outside Europe and North America and for animals whose instincts are unknown. *Natura non facit saltum* applies to instincts as well. Changes to instinct might be facilitated when one species exhibits different instincts during different periods of life, different seasons of the year, different circumstances, and so on, in which case one or the other instinct may be preserved by natural selection. Such examples of diverse instincts within one species do exist in the wild.

As is the case with bodily structures, and conforming with my theory, an instinct is good for its possessor and is never produced for the exclusive benefit of another species. A good example of an animal apparently performing an action solely for the good of another is aphids voluntarily yielding their sweet excretion to ants. To show that the action is voluntary, I removed all the ants from a group of about a dozen aphids on a dock plant. After several hours, the aphids wanted to excrete. I watched them through a magnifying lens, but none excreted. Then I tickled and stroked them with a hair, trying as best I could to imitate the action of ants' antennae, but none excreted. I finally allowed one ant to visit them; it seemed aware of what a rich flock it had discovered by its eager way of running about. Using its antennae, the ant began to play with the abdomen of one aphid, then another. As soon as each aphid felt the antennae, it immediately lifted its abdomen and excreted a clear drop of sweet juice, eagerly devoured by the ant. Even the young aphids acted this way, showing the action to be instinctive rather than gained by experience. Because the excretion is extremely sticky, it is probably convenient for the aphids to have it removed; they probably do not instinctively excrete it for the sole good of the ants. Although no animal in the world performs an action for the exclusive good of another species, each species tries to take advantage of other's instincts (just as each takes advantage of other's weaker bodily structures). So again, instincts are not absolutely perfect in some cases, but these and other details are dispensable and will here be passed over.

Natural selection requires the variation and inheritance of instincts, so as many examples as possible should be given here, but limited space prevents me. I can assert only that instincts *do* vary. For example, the migratory instinct varies in extent and direction and can be totally lost. Birds' nests also vary, partly in dependence on the place chosen and the nature and temperature of the inhabited region, but they also often vary from wholly unknown causes. Audubon has recorded several remarkable cases of differences in nests of the same species in the northern and southern United States. Fear of some particular predator is certainly an instinctive quality, as demonstrated by nestling birds, although it is strengthened by experience and the sight of fear of the same predator in other animals. Fear of humans is acquired slowly by various animals inhabiting desert islands, as I have shown elsewhere. There is an example of this even in England, where large birds are wilder than small birds because they have been most persecuted by humans. This must be the reason, because on uninhabited islands large birds are *not* more fearful than small birds, and the magpie, which is so wary in England, is tame in Norway, as is the hooded crow in Egypt.

Many observations show the diversity of general disposition among individuals of the same species. Also, several cases could be given of strange habits that might give rise to new instincts through natural selection, if advantageous to the species. I am aware that these general statements produce a feeble effect in the absence of detailed facts; I can only restate that I do not assert them without good evidence.

A brief consideration of examples from among domesticated animals will both strengthen the concept of heritable variations of instinct and demonstrate the respective roles of habit and selection of so-called accidental variations in modifying the mental qualities of domesticated animals. A number of curious and authenticated examples could be given of inherited shades of disposition and odd tricks associated with certain frames of mind or periods of time. But just consider the familiar example of certain dog breeds. Young pointers sometimes point and even back other dogs the very first time they are taken out (I saw a striking example myself). Retrieving is inherited to some degree by retrievers, and so is the sheepdog's tendency to run around sheep, rather than directly at them. These actions do not differ essentially from instincts; they are performed without experience by the young, nearly the same way by each individual, with eager delight by each breed, and without the end being known (the young pointer can no more know that he points to aid his master than the white butterfly knows why she lays her eggs on the leaf of cabbage). If the young of one kind of wolf were observed to stand motionless and then slowly crawl forward with a particular gait as soon as it scented prey, and if another kind of wolf were observed to rush around, instead of at, a herd of deer, then the actions of the two wolves would be called instinctive. Domestic instincts are far less fixed than instincts in the wild, but then they have faced less vigorous selection and have been transmitted for an incomparably shorter amount of time in a less constant environment.

Crossing different dog breeds shows how strongly these domestic instincts, habits, and dispositions are inherited and how curiously they become mingled. A cross with a bulldog has affected the courage and obstinacy of greyhounds, and a cross with a greyhound has given a whole family of sheepdogs a tendency to hunt hares. When these domestic instincts are thus tested by crossing, they resemble natural instincts, which also become curiously blended and exhibit traces of the instincts of either parent for a long time. For example, Le Roy describes a dog whose great-grandfather was a wolf. It showed only one trace of its wild parentage: it did not go to its master in a straight line when called.

It is not true that domestic instincts become heritable solely due to compulsory habit. No one would ever have thought of teaching the tumbler pigeon to tumble and probably would not have been able to even if he did. Young birds tumble despite never having witnessed another bird do so. At some point one pigeon showed a tendency for this strange habit; selection of the best tumbling individuals over the course of many generations made tumblers what they are now. Mr. Brent informs me that near Glasgow there are house tumblers that cannot fly eighteen inches high without going head over heels. No one would have thought to train a dog to point had not some one dog exhibited a natural tendency to do so. This happens occasionally, as I once observed in a purebred terrier. After the tendency is first displayed, methodical selection and the inherited effects of compulsory training in each generation complete the work. Unconscious selection is still at work as people try to procure dogs that stand and hunt best. In some cases habit alone has sufficed. No animal is more difficult to tame than the young of a wild rabbit, yet few animals are tamer than the young of a tamed rabbit. But domesticated rabbits have not been selected for tameness, and the entire change from extreme wildness to extreme tameness results simply from habit and long-term confinement.

Natural instincts are lost under domestication. A remarkable example is when fowl breeds rarely or never become "broody" – that is, they rarely sit on their eggs. Familiarity hinders our ability to recognize how universally and extensively domestication modifies the mind. Dogs instinctively love humans. Tame wolves, foxes, jackals, and species of the cat genus would eagerly attack poultry, sheep, and pigs. This tendency is incurable in dogs brought home as puppies from regions such as Tierra del Fuego and Australia, where the natives have not domesticated these animals. But domesticated European dogs rarely have to be taught not to attack poultry, sheep, and pigs even when they are young. They occasionally do make attacks but are subsequently chastised; if uncured they are destroyed. So habit and some degree of selection have probably acted together in domesticating these dogs. However, young chickens have lost wholly by habit the fear of cats and dogs, which was probably originally instinctive, as it is in young pheasants even when reared by a hen. Not that chickens have lost all fear, only fear of cats and dogs; if a hen gives the "danger" chuckle, her young run from under her and conceal themselves in the surrounding grass or thickets. This is evidently done instinctively to allow the mother to fly away, as observed in wild ground birds. Although this instinct has been retained by chickens, it has become useless because the hen has lost its ability to fly through disuse.

So domestic instincts are acquired and natural instincts lost partly by habit and partly by human selection and accumulation of peculiar mental habits that first appear by what we in our ignorance are forced to call "accident." In some cases compulsory habit suffices to produce heritable mental changes; in other cases compulsory habit does nothing, and selection – methodical and unconscious – determines the results. In most cases habit and selection probably act together.

I will use three cases to illustrate how instincts in the wild become modified by selection: (1) the instinct that leads the cuckoo to lay her eggs in other birds' nests, (2) the slave-making instinct of certain ants, and (3) the comb-making instinct of the honeybee. The latter two of these are generally and justly ranked as the most wonderful of all known instincts.

It is now commonly understood that the ultimate cause of the cuckoo's instinct is that she lays her eggs at intervals of two or three days rather than all at once. If she made her own nest and sat on her own eggs, then the first-laid eggs would have to be left unincubated for some time, or there would be eggs and hatchlings of different ages in the same nest. Laying and hatching would take inconveniently long, especially because cuckoos migrate early; the first-hatched young would probably have to be fed by the male alone. The American cuckoo is in this predicament: she has her own nest and cares for successively hatched nestlings simultaneously.[1] Suppose that the ancient ancestor of the European cuckoo had had the habits of the American cuckoo but occasionally laid an egg in another bird's nest. If it profited by this occasional habit, or if the young were more vigorous from taking advantage of the mistaken maternal instinct of another bird than from their own mother's care (encumbered as she would have been from having eggs and young of different ages at the same time), then the habit would have been advantageous. From inheritance, young birds reared this way would have perpetuated the aberrant habit of their mother and thus been successful in rearing *their* young. The strange instinct of the European cuckoo has been generated by a continuation of this process. In addition, according to Dr. J. E. Gray and some other observers, the European cuckoo has not completely lost all maternal love for her offspring.

The habit of birds laying eggs in other birds' nests (either of the same or distinct species) is common in the Gallinaceae, possibly explaining the origin of a unique instinct in the related ostrich group. At least in the case of the American species, multiple hen ostriches first lay a few eggs in one nest and then in another; these are hatched by the males. This instinct probably results from hens laying a large number of eggs and, like the cuckoo, at intervals of two or three days. However, the instinct is not

1. Nevertheless, there are several birds that occasionally lay eggs in other birds' nests.

perfected, and a surprising number of eggs are strewn over the plains; in one day's hunting I picked up no less than twenty lost and wasted eggs.

Many bees are parasitic and lay their eggs in the nests of other kinds of bees. This case is more remarkable than that of the cuckoo, because bees not only have their instincts modified in accordance with parasitic habits, but their structure is modified as well: they lack the pollen-collecting apparatus that would be necessary if they had to store food for their own young. Some *Sphegidae* (wasp-like insects) species are also parasitic. M. Fabre recently showed that *Tachytes nigra* generally makes its own burrow and stores it with paralyzed prey for its larvae to feed on, but if it finds a burrow made and stocked by another sphex, it takes advantage of the prize and becomes parasitic for the occasion. This case and that of the cuckoo show that natural selection can make an occasional habit permanent if it is advantageous to the species and, in the case of the wasp, if the insect whose nest and food are appropriated is not consequently exterminated.

The remarkable slave-making instinct was first discovered in the *Formica* (*Polyerges*) *rufescens* ant by Pierre Huber, a better observer than even his celebrated father. This species is completely dependent on its slaves and would go extinct in a single year without their aid. The males and fertile females do no work. The sterile female workers energetically and courageously capture slaves but do no other work. They cannot make their own nests or feed their own larvae. When a nest becomes old and inconvenient and they have to move, the slaves determine the migration and actually carry their masters in their jaws. The masters are so utterly helpless that when Huber shut thirty of them up without a slave but with plenty of their favorite food and with larvae and pupae to stimulate them to work, they did nothing. They could not even feed themselves, and many perished from hunger. He then introduced a single slave of the species *Formica fusca*, which instantly set to work. It fed and saved the survivors, made some cells and tended the larvae, and put everything right. Extraordinary! If no other slave-making ant were known, it would have been hopeless to speculate how so wonderful an instinct could have been perfected.

Formica sanguinea was also first discovered to be a slave-making ant by Pierre Huber. It is found in the southern parts of England, and its habits have been studied by Mr. F. Smith of the British Museum. Although fully trusting the statements of Pierre Huber and Mr. F. Smith, I tried to approach the subject in a skeptical frame of mind, as anyone may be excused for doing when studying an instinct as extraordinary and repulsive as slave making. I opened fourteen *Formica sanguinea* nests and found a few slaves in all of them. Males and fertile females of the slave species have never been observed in the nests of *Formica sanguinea*. The slaves are black-colored and less than half the size of their red-colored masters, so there is a great contrast of appearance. The slaves occasionally come out if a nest is slightly disturbed and defend it like their masters. When a nest is greatly disturbed, exposing larvae and pupae, the slaves work energetically with their masters in carrying them away to a safe place. It is clear that the slaves feel quite at home. During June and July of three successive years, I watched several nests in Surrey and Sussex for several hours and never saw a slave leave or enter a nest. The slaves are few during these months, so I thought they might behave differently when more numerous, but Mr. F. Smith informs me that he has watched nests at various hours during May, June, and August in Surrey and Hampshire, and although there are many slaves in August, he did not see any of them enter or leave a nest. He therefore considers them strictly "household" slaves, whereas the masters constantly bring in materials for the nest and all kinds of food. During July of this year, I found a community with an unusually large stock of slaves; I observed a few slaves mingled with their masters leaving the nest and marching to a tall Scotch fir tree twenty-five yards away, which they climbed together, probably in search of aphids and berries. According to Pierre Huber, in Switzerland the slaves habitually work with their masters in making the nest, they alone open and close the doors in the morning and evening, and their main job is to search for aphids. This difference in the usual habits of masters and slaves in the two countries probably results simply from the greater number of slaves being captured in Switzerland than in England.

One day I fortuitously witnessed a migration from one nest to another, and it was interesting to see the masters carefully carrying the slaves in their jaws, as Pierre Huber has described. Another day I noticed about twenty slave makers haunting the same spot and evidently not in search of food. They approached an independent community of the slave species *Formica fusca* and were vigorously repulsed. Sometimes as many as three clung to the legs of a slave-making *Formica sanguinea,* which ruthlessly killed their small opponents and carried their dead bodies as food to their nest, twenty-nine yards away. They were prevented from getting any pupae to rear as slaves. I then dug up a small parcel of *Formica fusca* pupae from another nest and put them down on a bare spot near the place of combat. They were seized and carried off by the tyrants, who maybe thought they had been victorious after all.

At the same time and place I laid a small parcel of *Formica flava* pupae with a few of these little ants still clinging to the nest fragments. As described by Mr. F. Smith, this species is rarely made into slaves. Despite its diminutive size it is very courageous, and I have seen it ferociously attack other ants. (I once found, to my surprise, an independent *Formica flava* community under a stone beneath a slave-making *Formica sanguinea* nest. When I accidentally disturbed both nests, the little ants attacked their big neighbors with surprising courage.) The *Formica sanguinea* immediately distinguished the pupae of *Formica fusca,* which they habitually make into slaves, from the pupae of the furious *Formica flava,* which they rarely capture; they instantly seized the *Formica fusca* pupae but were terrified when they came across the pupae or even the earth from the nest of the *Formica flava.* About fifteen minutes after the little yellow ants crawled away, the *Formica sanguinea* took heart and carried off their pupae.

One evening I visited another *Formica sanguinea* community and found ants carrying the dead bodies of *Formica fusca* (so it was not a migration) and pupae into the nest. I traced the booty-burdened returning file to a very thick clump of heath forty yards away and saw the last *Formica sanguinea* emerge carrying a pupa. I could not find the ravaged nest in the thick heath, but it must have been near, because two or three *Formica fusca* were rushing about in agitation, and one was perched motionless on the top of a spray of heath with its own pupa in its mouth.

Notice the great contrast between the instincts of *Formica sanguinea* and *Formica rufescens. Formica rufescens* does not build its own nest, determine its own migrations, collect food for itself and its young, or even feed itself: it is completely dependent on its numerous slaves. *Formica sanguinea* possesses fewer slaves – extremely few in the early part of the summer. The masters determine when and where a new nest will be formed, when to migrate, and the masters carry the slaves. Both in Switzerland and England the slaves seem to have exclusive care of the larvae, and only the masters go on slave-capturing expeditions. In Switzerland the slaves and masters work together in bringing materials for the nest and constructing it. Both tend the aphids, though this is done mostly by the slaves, and therefore both collect food for the community. In England the masters leave the nest to collect building materials and food for themselves, their slaves, and their larvae. The master ants in England receive less service from slaves than the master ants in Switzerland.

Ants that are not slave makers sometimes carry off other species' pupae if they are scattered near the nest, and pupae originally stored as food might become developed. Ants unintentionally reared this way follow their instincts and do what work they can. If their presence proves useful to the species that seized them, and if it is advantageous to capture workers rather than procreate them, then the habit of capturing pupae for food might be strengthened by natural selection and rendered permanent for the very different purpose of raising slaves. Once the instinct is acquired, at this stage being less developed than even in the British *Formica sanguinea* (which is aided less by its slaves than the same species in Switzerland), it can be modified by natural selection until an ant results as abjectly dependent on its slaves as *Formica rufescens* so long as each modification is useful to the species.

I will not discuss the details of the honeybee's cell-making instinct but will simply give an outline of my conclusions. Only a dull person can examine the exquisite structure of a honeycomb, so perfectly adapted to its end, without enthusiastic admiration. We hear from mathematicians that bees have practically solved an abstruse problem and make their cells the right shape to hold the maximum amount of honey with the minimum amount of precious wax. A skillful worker equipped with the right tools would find it difficult to construct well-formed wax cells, but a crowd of bees in a dark hive construct them perfectly. Even granting any imaginable instinct, it seems at first inconceivable that bees can determine all the necessary angles and planes, or even perceive when they are correctly made. But the difficulty is not nearly as great as it seems, and all this beautiful work follows from a few very simple instincts.

Mr. Waterhouse inspired me to investigate this subject. He has shown that the form of each cell is defined by adjoining cells, and the following is a modification of his theory. Using the principle of gradation, let us see if Nature reveals her method. At one end of a short series are bumblebees, which use their old cocoons, sometimes with wax tubes, to hold honey and also to make separate irregular and rounded wax cells. At the other end of the series are honeybees. Each cell of the comb is a hexagonal prism with slanted basal edges that form a pyramid. The cells are made in a double layer, with the openings facing opposite directions. The pyramidal base of a cell in one layer of the comb therefore partially defines the shape of the bases of three adjoining cells on the opposite layer. In between the bumblebee's simple cells and the honeybee's perfect cells are those of the Mexican *Melipona domestica*, carefully described and depicted by Pierre Huber. The bee itself is structurally intermediate between the bumblebee and honeybee, but more closely related to the bumblebee. It constructs both a nearly regular wax comb of cylindrical cells in which the young are hatched, and an irregular mass of nearly spherical, nearly equal-size, large cells for honey. Importantly, if the spheres were perfect, they would intersect or break into one another, but instead the bees build flat wax walls between spheres. So each cell is roughly spherical with two, three, or more perfectly flat surfaces, depending on the number of adjoining cells. If one cell contacts three others – and this happens frequently, because each sphere is nearly the same size – then the three flat surfaces are united into a pyramid. And as Huber has remarked, these pyramids are roughly similar to the three-sided pyramidal bases of honeybee cells. As with the honeybee's cells, the three-plane surface of any one *Melipona domestica* cell partially defines the shapes of three adjoining cells. In this way *Melipona domestica* obviously saves wax, because each flat wall between two cells is the same thickness as the rest of the spherical cell (i.e., it is not doubly thick), but it forms a part of *two* cells.

   While reflecting on this case it occurred to me that if *Melipona domestica* were to make equal-size spheres at some set distance from each another and arrange them symmetrically in a double layer, the resulting structure would be as perfect as the comb of the honeybee.[2] Only slight

---

2. I wrote to the mathematics Professor Miller of Cambridge, and he kindly read over the following, drawn up from his information, and tells me that it is correct.

modifications in the instincts of *Melipona domestica,* which are not particularly wonderful, would be required. *Melipona domestica* would have
to make truly spherical and equal-size cells, but it already does this to a
certain extent. (Many insects can make perfectly cylindrical burrows in
wood, apparently by turning on a fixed point.) It would have to arrange
cells in layers, as it already does with the cylindrical cells, and each bee
would have to accurately judge what distance to stand from her fellow
workers when multiple spheres were being constructed simultaneously.
Melipona domestica can already judge distance, because spheres are
constructed to intersect, and then the points of intersection are united
by perfectly flat surfaces. Furthermore, once the hexagonal prisms are
formed by intersecting spheres in the same layer, they would have to be
able to prolong the hexagon to any length in order to hold the stock of
honey (similarly to the crude bumblebee, which adds cylinders of wax
to the circular mouths of spent cocoons). Such modifications of simple
instincts – hardly more complex than the instincts that guide a bird to
construct its nest – through natural selection have given the honeybee
its inimitable architectural powers.

This theory can be experimentally tested. Following Mr. Tegetmeier's example, I separated two combs and put a long, thick, square strip of
wax between them. The bees instantly began excavating minute circular
pits, which became wider and wider until they were shallow basins with
circular openings that were about the diameter of a cell. Interestingly,
whenever several bees excavated basins near one another, they began this
distance apart. When the basins reached a depth of about one-sixth the
diameter of the sphere of which they formed a part, the rims of the basins
intersected and broke into one another. As soon as this happened, the
bees stopped excavating and began building flat wax walls on the lines
of intersection so that each hexagonal prism was built on the edge of a

---

Consider two parallel layers of equal-size spheres with the center of each sphere
(radius $\times \sqrt{2}$) units away from the center of the six spheres surrounding it in the same
plane as well as the adjoining spheres in the second plane. If planes of intersection are
described between the spheres of both layers, the result is a double layer of hexagonal
prisms united by pyramidal bases. The edges of the hexagonal prisms form angles identical to the best available measurements of the corresponding angles in honeybee cells.

smooth basin instead of on the straight edges of a three-sided pyramid like ordinary cells.

Next, I put a thin and narrow strip of wax into the hive. The bees began excavating little basins on both sides as before. But the wax was so thin that if the basins had been excavated as deeply as before, their bottoms would have broken into one another from opposite sides. The bees did not suffer this to happen and stopped excavating in time. As soon as the basins were a little deepened, their bottoms were flattened. These thin plates of un-gnawed wax were situated along the imaginary planes of intersection between opposite basins and were unequal because the work was not neat (the conditions being unnatural). The bees on the two sides of the strip must have worked at nearly the same rate to leave flat plates by stopping work along the planes of intersection. Because wax is so flexible, the bees easily perceived when they had gnawed it to the proper thinness. In ordinary combs the bees do not always work at exactly the same rate from opposite sides, and I have seen half-completed common walls at the base of a just-started cell that were slightly concave on one side (where the bees excavated too quickly) and convex on the other (where the bees excavated too slowly). In one case I put the comb back into the hive and allowed the bees to go on working for a short time; on reexamination I found that the common wall had been completed and rendered perfectly flat! The bees could not have possibly done this by gnawing away at the convex side, because the plates of wax were so thin. I suspect that in such cases the bees stand in opposite cells and push and bend the warm wax into its proper plane, thereby flattening it.

This experiment demonstrates that if bees were to build themselves a thin sheet of wax, they could make their properly shaped cells by standing at the right distance from one another, excavating at the same rate, and making equal-size spherical hollows that are never allowed to break into one another. In fact, an inspection of the edge of a growing comb shows that bees make a rough circumferential rim all around the comb; they gnaw into it from opposite sides and work circularly as they deepen each cell. They do not make the entire three-sided pyramidal base of any one cell at the same time, only one or two sides of the base, and they never complete the upper edges of these bases until the hexagonal walls are started. Some of these assertions differ from those made by the justly celebrated elder Huber, but I am convinced of their accuracy and I could show they agree with my theory if I had more space. (Huber's statement that the first cell is excavated out of a little parallel-sided wax wall is not strictly correct, the first excavation always being a little hood of wax. I will not enter into details here.)

Although excavation plays an important part in the construction of cells, it would be incorrect to assume that bees cannot build a rough wax wall in the right position – that is, along the plane of intersection between two spheres. I have several specimens clearly showing that they can do this. Curved parts that correspond in position to the future basal planes of cells sometimes occur even in the crude circumferential rim of wax around a growing comb. But the rough wall is always finished off by being gnawed away on both sides. The bees have a curious way of building: they make the first rough wall ten to twenty times thicker than the thin wall of a finished cell. By comparison, imagine a group of masons who make a thick cement wall. They then begin cutting it away equally from both sides near the ground until a thin wall is left in the middle. The masons remove the cut-away cement but keep adding fresh cement to the top of the wall as the cutting away moves upward. The result is a thin wall growing upward but always covered by a huge crown of cement. In the case of bees, this strong crown of wax over all the cells (both completed and just begun) allows the bees to cluster and crawl over the comb without injuring the delicate hexagonal walls, which are only about four-hundredths of an inch thick (the pyramidal bases are only about twice as thick). This singular mode of construction strengthens the comb and economizes on wax.

A multitude of bees work together to build cells, seemingly making our understanding of the process more difficult. One bee works on a cell for a short time and then goes to another; as Huber has stated, twenty individual bees work at starting even the first cell. I demonstrated this by covering the edges of a single cell's hexagonal walls or the margin of the circumferential rim with a thin layer of red wax; the color was as delicately diffused by the bees as a painter could have done with his brush. Minuscule amounts of the colored wax were taken from where it was placed and worked into the growing edges of all the cells. Construction work seems to be a sort of balance struck by many bees all instinctively standing at the same relative distance from one another while sweeping out equal spheres, and then building up or leaving un-gnawed the planes of intersection between these spheres. (It is really interesting that in a difficult situation, like when two pieces of comb meet at an angle, the bees entirely pull down and rebuild the same cell in different ways, sometimes reverting to a shape that was at first rejected.)

When bees encounter a place where they can stand properly, like a slip of wood directly beneath the middle of a downward-growing comb that has to be built over one face of the slip of wood, they can lay the foundations of one wall of a new hexagon projecting beyond the other completed cells. It would be sufficient for the bees to stand at their correct relative distance from one another and from the walls of the last completed cells and then, by striking imaginary spheres, build up a wall intermediate between two adjoining spheres. But as far as I have seen, they never finish off the angles of a cell until a large part of both that cell and the adjoining cell has been built. (This capacity to lay down a rough wall in its correct position in certain circumstances is important because it bears on an observation that at first seems subversive to my theory – namely, that the cells on the margins of wasp combs are sometimes perfectly hexagonal. But I have no space to discuss this topic.) Even a single insect, like the queen wasp, can make hexagonal cells if she alternately works on the insides and outsides of two or three cells. She stands at the correct relative distance between cells and sweeps out spheres or cylinders and builds up intermediate planes. It is even conceivable that an insect can make an isolated hexagon by choosing a fixed point and then moving out first to one point, then five others at the correct relative distance from the fixed point before striking the planes of intersection. However, I do not know of any such case, nor would any good be derived from building an isolated hexagon, because more material would be required than in the construction of a cylinder.

Because natural selection accumulates slight modifications of instinct, each profitable to the individual under specific environments, it is reasonable to consider how a long and graduated series of cell-building instincts culminating in the current system of construction could have profited the honeybee's ancestors. This is not difficult. Bees are often hard-pressed to get enough nectar. Mr. Tegetmeier informs me that a

hive of honeybees consumes twelve to fifteen pounds of sugar for the secretion of each pound of wax, so the construction of wax combs requires
the collection and consumption of prodigious amounts of nectar. Also,
many bees have to remain idle for days to secrete wax. A large store of
honey is indispensable to the support of a large stock of bees during the
winter, and a hive's security depends mainly on a large number of bees.
Thus, economical use of wax in storing copious amounts of honey must
be an important element of success for any bee family. Of course the
success of any bee species may depend on the number of its parasites
or other enemies, or on other factors, and be altogether independent
of how much honey the bees can collect. But suppose that the effective
storage of honey determines (as it often probably does determine) the
number of bumblebees that exist in a region. Further suppose that the
community lives through the winter and consequently requires a honey
cache. In this case a slight modification of instinct that leads to cells
being nearer one another would definitely be an advantage, because if
each cell has even one wall in common with another, then some wax is
saved. It would be even more advantageous if the bumblebee made cells
more regular, closer together, and increasingly aggregated, like the cells
made by *Melipona domestica*. The greater the surface of each cell shared
in common with others, the less wax is required. Similarly, it would be
advantageous for *Melipona domestica* to make its cells more regular and
closer together than it makes them now; the spherical surfaces would
become planar surfaces and *Melipona domestica* would make a comb
as perfect as the honeybee's. Natural selection cannot lead beyond this
stage of perfection, because the honeybee comb is absolutely perfect in
economizing on wax.

So the most wonderful of all instincts can be explained by natural
selection having taken advantage of many successive modifications of
simpler instincts. By slow degrees, natural selection led bees to sweep
out equal spheres at a set distance from one another in a double layer
and to build up and excavate the wax along the planes of intersection.
Of course the bees comprehend no more that they sweep out spheres at
a particular distance than they know the angles found in the hexagonal
prisms and pyramidal bases. With the motive power of natural selection being economy of wax, the individual swarms that succeeded and

transmitted their new economical instinct and had the best chance of succeeding in the struggle for existence were those wasting the least honey in secreting wax.

No doubt many difficult-to-explain instincts could be cited against the theory of natural selection: cases in which the instinct's origin cannot be explained, no intermediate gradations are known to exist, the instinct seems so unimportant that natural selection could hardly have acted on it, or the instinct is almost identical to that of a very remotely related animal so that inheritance from common ancestry is unlikely and independent acts of natural selection are likely. I will not discuss all of these cases here, but I will address one natural occurrence that may seem insurmountable and actually fatal to my theory: neuter (sterile) females in insect communities. These neuters are often structurally and instinctually very different from both the males and fertile females and yet, being sterile, cannot propagate their kind.

Although this subject should be explored at length, I will consider only the example of sterile worker ants. How the workers have been rendered sterile is difficult to explain, but not more so than any other striking structural modification. Insects and other articulated animals occasionally become sterile in the wild, and if the organism is social, it is profitable to the community that a few individuals be capable of work but incapable of reproduction. Natural selection can effect this. The great difficulty lies in the significant difference between worker ants and both the males and fertile females – for example, in thoracic structure, the absence of wings and sometimes eyes, and instinct. If the worker ant were not neuter, I would unhesitatingly assume that all of its characteristics had been acquired through natural selection, because an individual with a slight profitable modification passes it on to offspring; these vary and are selected, and so on. But the worker ant is very different from its parents, yet sterile, so it cannot transmit successively acquired modifications to progeny. How can this be reconciled with the theory of natural selection?

First, recall that there are many domestic organisms and organisms in the wild with all sorts of structural differences correlated with certain ages or with either sex. There are differences correlated not only with one sex but also with just the short period when the reproductive system is active. (Examples include the nuptial plumage of many birds and the hooked jaws of male salmon.) There are even slight differences in the horns of different cattle breeds due to an artificial imperfection of males – for example, oxen of certain breeds have longer horns than bulls or cows of other breeds. So there is no real difficulty in a characteristic becoming correlated with sterility in certain members of some insect communities; the difficulty is in understanding how such correlated structural modifications could have been slowly accumulated by natural selection.

This difficulty appears insurmountable, but it actually disappears when one considers that natural selection can be applied to the family as well as the individual. Although a tasty vegetable is destroyed when it is cooked, the gardener sows seeds of the same stock and confidently expects nearly the same variety. Cattle breeders want flesh and fat marbled, so if an individual cow is slaughtered and yields such meat, the breeder goes confidently to the same family of cows. I have such faith in the power of selection that I believe a breed of cattle that always yields oxen with long horns could be slowly formed by carefully tracking which bulls and cows, when paired, produce oxen with the longest horns. A similar process has played out with social insects. A slight structural or instinctual modification correlated with sterility is advantageous to the community, allowing the fertile male and female members of the community to flourish and transmit to their fertile offspring the capacity to produce sterile offspring, along with the same modification. A repetition of this sequence produced the huge difference between sterile and fertile insects observed in certain species today.

The climax of the difficulty has not been touched on yet – namely, that the neuters of several ant species differ not only from the fertile males and females but also from each other, sometimes to an incredible degree, and are divided into two or even three castes. Furthermore, the castes do not usually graduate into one another, each one being as distinct from the others as a species is from members of its genus or a genus from members of its family. In *Eciton* there are worker and soldier neuters with extraordinarily different jaws and instincts. In *Cryptocerus* the workers of one caste carry a wonderful sort of shield on their heads with an unknown function. In the Mexican *Myrmecocystus* the workers of one caste never leave the nest, are fed by the workers of another caste, and develop a huge abdomen that secretes a sort of honey.

Some readers might think I am overconfident in natural selection when I do not admit that such wonderful established facts annihilate my theory. Neuter insects that are all the same (i.e., when there is only "one caste") can be rendered different from fertile males and females by natural selection. By analogy with ordinary variation, each slight, successive, and profitable variation did not appear in all the individual neuters simultaneously, but only in a few. All the neuters ultimately come to possess the desired characteristic by the selection in each generation of the fertile parents that produced the most neuters with the profitable modification. This being true, there should occasionally be neuter insects of the same species and in the same nest exhibiting structural gradations – and there are (quite often, in fact, considering how few neuter insects have been fully examined in Europe). Mr. F. Smith has shown that neuters of several British ants differ surprisingly in size and sometimes color and that the extreme forms can sometimes be linked together by individuals from the same nest (I have observed such perfect gradations myself). The large- or small-size workers (or both) are often numerous relative to the intermediate workers. In *Formica flava*, which has small, large, and some intermediate workers, Mr. F. Smith has observed that the large workers have simple eyes – small "ocelli" that can be easily distinguished – whereas the small workers have rudimentary ocelli. I carefully dissected several of these workers and can affirm that the ocelli of small workers are far more rudimentary than can be accounted for by their lesser size; the intermediate workers have ocelli in an intermediate condition. So in this case one nest contains two sets of sterile ants differing in size and eye structure yet connected by a few individuals in an intermediate condition. (If the small workers had been more useful to the community, the males and females that produced the greatest number of small workers would have been selected until all the workers had become small. The result would be an ant species with neuters similar to those of *Myrmica*, which do not even have rudimentary ocelli, though the male and female ants of the genus have well-developed ocelli.)

So confidently did I expect to find gradations of structure between different neuter castes of the same species that I gladly accepted Mr. F. Smith's offer of West African driver ant specimens. In order to appreciate the amount of difference between these workers, I will give an accurate parallel. Imagine a group of workmen building a house; many of them are five feet four inches tall, and many are sixteen feet tall. Also, the larger workmen have heads that are four times as big as those of the small workmen and jaws that are nearly five times as big. The jaws and the form and number of teeth of the various-size worker ants differ wonderfully. Although the workers can be grouped into castes based on size, the sizes graduate into one another, as do the different jaw structures. I am confident about this last point because Mr. Lubbock used a camera lucida to draw the jaws I dissected from workers of several sizes.

These considerations taken together show that by acting on fertile parents, natural selection can form a species that regularly produces neuters. These neuters can be (1) large with one jaw form, (2) small with some other jaw form, or, most crucially, (3) of one set of workers of one structure and size and another set of a different structure and size. This third scenario is preceded by the formation of a graduated series of workers, as found in the driver ant. Then the extreme forms are produced in greater and greater numbers, because they are most useful to the community, through the natural selection of the parents that generated them, and eventually no workers with an intermediate form are left.

The development of two sterile castes is useful to a social insect community in the same way that division of labor is useful to humans. Ants inherit instincts, tools, and weapons and cannot acquire knowledge or manufacture instruments, so a perfect division of labor can only be effected if the workers are sterile; if they were fertile they would cross and blend their instincts and structure. Nature has effected such a division of labor through natural selection. I confess that despite all my faith in this principle, I would never have anticipated natural selection to be so efficient had the case of neuter insects not convinced me. I have discussed this example to demonstrate natural selection's power and because it is by far the most serious specific challenge to my theory. It is also very interesting because it shows that with animals, as with plants, the accumulation of many small, "accidental," and profitable variations can effect any amount of modification without exercise or habit. No amount of exercise, habit, or volition on the part of sterile insects could possibly influence the structure or instincts of the fertile community members that leave the offspring. I am surprised no one has advanced the case of neuter instincts against Lamarck's well-known doctrine.

In this chapter I have briefly shown that the mental qualities of domestic animals vary and that these variations are inherited. Even more briefly I have shown that instincts vary in the wild (that instincts are very important to every animal is not disputed); therefore, as the environment changes, natural selection can accumulate slight modifications of instinct to any extent in any useful direction. Habit or use and disuse are probably relevant in some cases. The observations of this chapter do not necessarily strengthen my theory, but none of the difficult cases annihilates it either. However, the observations that few instincts are perfect, no instinct is for the exclusive good of another animal, each animal takes advantage of other's instincts, and that *natura non facit saltum* applies to instincts and can be clearly explained by the above concepts but is otherwise inexplicable, all corroborate the theory of natural selection.

The theory is also strengthened by other observations about instinct. Closely related but distinct species commonly have nearly the same instincts even when they live in considerably different environments in distant parts of the world. For example, the principle of inheritance explains why the South American thrush lines its nest with mud in the same peculiar manner as the British thrush. Similarly, the male wrens of North America build "cock nests" to roost in, like the males of the distinct kitty wrens, a habit wholly unlike that of any other known bird. Finally, it may not be a logical deduction, but to my imagination it is far more satisfactory to look at instincts like the young cuckoo ejecting its foster siblings, ants making slaves, and ichneumonidae feeding within the living bodies of caterpillars as consequences of one general law than as specially endowed or created. This law leads to the "advancement" of each organic being: multiplication, variation, survival of the strong, and death of the weak.

8

HYBRIDS

NATURALISTS GENERALLY MAINTAIN THAT THE OFFSPRING
of crosses from between different species are specially endowed with sterility to prevent the confusion of all organisms. At first this seems probable, because species within a region would hardly remain distinct if they
were to cross freely. The importance of the fact that hybrids are generally
sterile has recently been underrated by some writers. But for the theory
of natural selection, hybrid sterility is a particularly important problem,
because sterility cannot possibly be useful to a hybrid and therefore cannot be acquired by the preservation of successive degrees of profitable
sterility. However, I hope to show that sterility is not a specially acquired
or endowed quality, but a by-product of other acquired qualities.

Two sets of observation, to a large extent fundamentally different,
are often confused in discussions of this subject: the sterility of two species when crossed, and the sterility of hybrids produced from a cross of
two species.[1]

Of course the reproductive organs of pure species are in perfect order, yet when two species are crossed they produce few or no offspring.
Hybrids, on the other hand, have functionally impotent reproductive

1. [Darwin is indicating the inability of most paired species to produce offspring,
referred to as "species sterility," or "sterility (or fertility) in the first cross." For example,
a cat and a dog are sterile with respect to each other, because even if a cat and a dog
mate, no offspring result. If two closely related species *can* produce offspring (i.e., there
is no species sterility), then these offspring are "hybrids," and they are often sterile:
"hybrid sterility" or "sterility (or fertility) of hybrids." Sometimes Darwin uses the word
"sterility" in a general sense to indicate reduced fertility. – D.D.]

organs (as is clearly observable in the male element of both plant and animal hybrids), even though the organs are structurally perfect as far as the microscope reveals. In the case of species sterility the two sexual elements that go on to form the embryo are each perfect, and in the case of hybrid sterility they are imperfectly developed or not developed at all. This distinction is important to consider in determining the cause of sterility; it has been glossed over because in both cases sterility is considered a special endowment beyond the explanation of our reasoning powers. The fertility of varieties when intercrossed and the fertility of their mongrel offspring are as important a consideration as species sterility, because it seems to make a clear and broad distinction between "variety" and "species."

First, consider species sterility and hybrid sterility. Two diligent and admirable observers, Kölreuter and Gärtner, devoted almost their entire lives to this subject. Their memoirs and works deeply impress the generality of some degree of sterility. Kölreuter makes the rule universal. But he uses a sleight of hand: he finds ten cases where two forms were considered distinct species but were nevertheless fertile together, so he unhesitatingly ranks them as varieties. Gärtner makes the rule equally universal by disputing the fertility of Kölreuter's ten cases. However, in these and in many other cases Gärtner is obliged to carefully count seeds in order to grant that there is any degree of sterility. He always compares the maximum number of seeds produced by two species when crossed, and those produced by their hybrid offspring, with the average number of seeds produced by both pure parent species in the wild. But this introduces a serious source of error: a plant that is to be hybridized must be castrated and, often more importantly, secluded in order to prevent pollen being brought to it by insects from other plants. Nearly all of the plants used in Gärtner's experiments were potted and apparently kept in a chamber of his house. Both of these conditions often reduce plant fertility: in his table of results, Gärtner lists about twenty plants that he castrated and artificially fertilized with their own pollen (excluding plants such as legumes, which are difficult to manipulate), and half of them had some degree of impaired fertility. Even more telling, over several years Gärtner repeatedly crossed the primrose and cowslip, which are acknowledged varieties, and obtained fertile seeds only once or twice.

He also found the common red and blue pimpernels absolutely sterile together, even though the best botanists rank them as varieties, and he comes to the same conclusion in several analogous cases. So it seems to me that it is fair to doubt whether many other species are really so sterile when intercrossed as Gärtner believes.

Species sterility varies so much, and the fertility of pure species is so easily affected by various circumstances, that for all practical purposes it is difficult to establish where perfect fertility ends and partial sterility begins. The best evidence for this is that the two most experienced observers who have ever lived – Kölreuter and Gärtner – arrive at diametrically opposite conclusions for the same cross! (It is also instructive to compare evidence advanced by botanists concerning whether doubtful forms should be ranked as species or varieties, with evidence about fertility adduced by hybridizers, or by one author from experiments conducted in different years.) Thus, it can be shown that neither sterility nor fertility affords a clear distinction between species and varieties, and the evidence from this source is as doubtful as that derived from other constitutional and structural differences.

In studying hybrid sterility through successive generations, Gärtner reared some hybrids for six, seven, or in one case ten generations, while carefully guarding them from crosses with the pure parent strains. He asserts that fertility never increased, and usually decreased significantly. It is probably true in general that fertility decreases suddenly in the first few generations. But I believe that in all of Gärtner's experiments fertility is diminished by an independent cause: inbreeding. I have collected so many observations that show that close inbreeding *reduces* fertility, and that occasional crossing with distinct individuals or varieties *increases* fertility, that these two almost universal beliefs of breeders are beyond question. Experimentalists rarely raise many hybrids, and because the parent species or other related hybrids usually grow in the same garden, the visits of insects must be prevented during the flowering season in order to inhibit cross-pollination. This means that with each generation the hybrids are generally fertilized by their own pollen. This impairs their fertility, already lessened by their hybrid origin. Gärtner repeatedly makes a remarkable statement that supports this conviction: if even the least fertile hybrids are artificially fertilized with pollen from another individual of the same hybrid kind, then their fertility sometimes increases despite the usual negative effects of manipulation. As I know from my own experience, during artificial pollination, pollen is often accidentally taken from the anthers of another flower as well as from the very flower that is to be fertilized. This produces an accidental cross between two flowers (albeit, usually from the same plant). To control this, whenever an experiment is complex, an observer as careful as Gärtner castrates the hybrids to ensure that in each generation a cross is between distinct flowers (whether on the same plant or on different plants). So the strange observation that fertility increases with each successive generation if artificial fertilization is used can be explained by the fact that carefully performed artificial fertilization prevents inbreeding.

Now consider the results collected by the third most experienced hybridizer, the Honorable and Reverend W. Herbert. He maintains that some hybrids are perfectly fertile – as fertile as their pure parent species – just as emphatically as Kölreuter and Gärtner maintain that offspring from crosses between different species are always somewhat sterile. (He experimented with some of the very same species as Gärtner.) The difference in their results may in part be explained by Herbert's great horticultural skill and his access to greenhouses. Of his many important statements, I will give just one as an example: "Every ovule in a pod of *Crinum capense* fertilized by *C. revolutum* produced a plant which I never saw to occur in a case of its natural fecundation." This is a case of perfect, even uncommonly perfect, fertility in the offspring of the first cross between two distinct species.

This example leads me to cite a singular observation: some plants are more easily fertilized by pollen from other species than by the pollen they produce themselves. (Examples include certain species of *Lobelia* and all members of the genus *Hippeastrum*.) These plants yield seed to the pollen of a distinct species but are sterile to the pollen they produce themselves – yet their own pollen is perfectly good because it can fertilize other species. So certain individual plants, and all the individuals of certain species, actually hybridize more readily than they self-fertilize! For example, a bulb of *Hippeastrum aulicum* produced four flowers. Herbert fertilized three of these with their own pollen and the fourth with the pollen of a compound hybrid descended from three other species.

The result: "The ovaries of the three first flowers soon ceased to grow, and after a few days perished entirely, whereas the pod impregnated by the pollen of the hybrid made vigorous growth and rapid progress to maturity, and bore good seed which vegetated freely." In a letter to me in 1839, he wrote that he had repeated the experiment over the course of more than five years and always got the same result. This finding has been confirmed by other observers for *Hippeastrum* and its subgenera, and for some other genera, such as *Lobelia, Passiflora,* and *Verbascum.* The plants in these experiments appeared perfectly healthy. Ovules and pollen from the same flower were perfectly functional with respect to the pollen and ovules of other species but could not respond to each other. This suggests that the plants must have been in an unnatural state. Yet these observations show how the greater or lesser fertility of a species when crossed sometimes depends on minor and mysterious causes, especially when compared to the fertility of the same species when it self-fertilizes.

Although the practical experiments of plant breeders are not made with scientific precision, they deserve some notice. Species of *Pelargonium, Fuchsia, Calceolaria, Petunia, Rhododendron,* and others have been crossed in exceptionally complicated ways, and yet these hybrids seed freely. For example, Herbert asserts that a cross between *Calceolaria integrifolia* and *Calceolaria plantaginea* gave rise to a hybrid that "reproduced itself as perfectly as if it had been a natural species from the mountains of Chile," even though the two parent species are very different. I have taken some pains to ascertain the degree of fertility of some of the complex crosses of rhododendrons, and I am assured that many of them are perfectly fertile. For example, Mr. C. Noble raises a stock of *Rhododendron ponticum–Rhododendron catawbiense* hybrids for grafting, and they "seed as freely as it is possible to imagine." If well-treated hybrids decreased in fertility with each generation as Gärtner believes, plant breeders would have noticed. Plant breeders raise large beds of the same hybrids, and individuals cross freely through insect pollination, so the negative influence of close inbreeding is avoided. The great efficiency of insects in transporting pollen is apparent on examining the flowers of certain very sterile hybrid rhododendrons; these produce no pollen themselves and yet have stigmas with plenty of pollen brought from other flowers.

Far fewer experiments have been carefully carried out with animals than with plants. If our systematic arrangements can be trusted (i.e., if animal genera are as distinct from one another as plant genera), then animals that are widely separated on the scale of nature can be crossed more easily than is the case with plants, although the hybrids themselves are more sterile. I doubt there is a thoroughly authenticated case of a perfectly fertile hybrid animal. However, because few animals breed freely in confinement, few experiments have been tried. For example, the canary has been crossed with nine other finches, but none of these nine species breeds freely in confinement, so there is no reason to expect perfect fertility in the first cross or of their hybrids. Concerning the fertility of the more fertile hybrid animals through successive generations, I know of only a few cases in which two families of the same hybrid have been raised at the same time from *different* parents in order to avoid the negative effects of inbreeding. In fact, brothers and sisters are usually crossed in each successive generation despite the constantly repeated admonitions of every breeder. So it's not at all surprising that the inherent sterility of the hybrids should continue increasing; if brothers and sisters of any pure animal were paired and had some tendency to sterility (from whatever cause), the breed would definitely disappear in a few generations.

Although I do not know of any thoroughly authenticated cases of perfectly fertile hybrid animals, there is evidence that hybrids of *Cervulus vaginalis* and *Cervulus reevesii,* and of *Phasianus colchicus* and both *Phasianus torquatus* and *Phasianus versicolor* are perfectly fertile. Even though the common goose and Chinese goose are such different animals that they are generally ranked in separate genera, their hybrid often breeds with either pure parent, and in one case a hybrid bred with a hybrid. This was effected by Mr. Eyton, who raised two hybrids from the same parents but different hatches. From these he raised at least eight hybrids – grandchildren of the pure geese – from one nest. In India, however, these cross-bred geese must be much more fertile; Mr. Blyth and Captain Hutton assure me that whole flocks of hybrids are kept in various parts of the country, and because they are kept for profit where neither pure parent species exists, they must be very fertile.

Modern naturalists largely accept the doctrine proposed by Simon Pallas that most domestic animals have descended from two or more original species, which have since then become commingled by crossing. If this is true, then either the original species immediately produced fertile hybrids, or the hybrids became fertile during subsequent generations under domestication. To me, the second alternative seems more likely, even though it rests on no direct evidence. For example, I believe that dogs have descended from several wild stocks, and yet all are fertile together (perhaps with the exception of certain domestic dogs indigenous to South America). By analogy, it is very unlikely that the original species would have bred together freely at first and produced quite fertile hybrids. Similarly, European and humped Indian cattle are fertile together, though based on facts supplied by Mr. Blyth, they are distinct species. This view of the origin of many domestic animals requires either that we give up the notion that crosses between species are almost universally sterile, or that we recognize that sterility can be removed by domestication.

These observations about crosses in plants and in animals lead to the conclusion that there is generally some degree of sterility in first crosses and of hybrids, but given our present state of knowledge of the subject, this cannot be considered absolutely universal.

The circumstances and rules governing sterility in first crosses and of hybrids need to be considered in more detail to discover whether or not species have been specially endowed with sterility to prevent their crossing and blending to utter confusion. The following conclusions are drawn up from Gärtner's admirable work on the hybridization of plants. I have attempted to discover how far the rules also apply to animals, and given our scanty knowledge of hybrid animals, I am surprised to find how generally the same rules apply to both kingdoms.

Fertility in first crosses and of hybrids graduates from zero to perfect. The many curious ways this gradation can be shown to exist is surprising, but only the barest outline of the facts can be given here. When pollen from a plant of one family is placed on the stigma of a plant from another family, it exerts no more influence than so much inorganic dust:

absolute zero fertility. Pollen from different members of the same genus used on some one species yields a perfect gradation in the number of seeds produced, sometimes up to quite complete fertility (and as already mentioned, in some abnormal cases up to excess fertility beyond that produced by the plant's own pollen). Now turning to the hybrids, there are some that have never produced a single fertile seed, even with pollen from either pure parent. But in some of these cases a trace of fertility can be detected when the pollen of a pure parent species causes a flower on the hybrid to wither earlier than usual (the early withering of a flower is a sign of incipient fertilization). From this extreme degree of sterility, the gradation moves through self-fertilizing hybrids that produce more and more seeds, up to perfect fertility.

Hybrids springing from two species that are very difficult to cross tend to be very sterile, but this correlation between the difficulty of making a first cross and the sterility of the resulting hybrid (two concepts that are commonly confused) is not strict. There are many pairs of species that can be easily crossed to produce hybrid offspring, and yet these hybrids turn out to be remarkably sterile. There are also pairs of species that can be crossed only with difficulty and yet produce very fertile hybrids. These two opposite cases even occur within the confines of one genus (*Dianthus*, for example).

Unfavorable conditions affect fertility in first crosses and of hybrids more easily than they affect the fertility of pure species. But fertility varies innately in all cases and isn't always the same in first crosses, even when the same two species are crossed under the same circumstances. It depends partly on the individuals chosen for the experiment. Similarly, hybrids raised from seeds out of the same capsule and under identical conditions often vary greatly in fertility.

"Systematic affinity" means the structural and constitutional resemblance between species, especially in the structure of parts that are very important physiologically and that differ little among related species. Now, fertility in first crosses and of hybrids is largely governed by the systematic affinity of the species in question. Species ranked by systematists in different families have never generated hybrids, whereas closely related species generally produce hybrids easily. But the correspondence between systematic affinity and the facility of crossing is not at all strict.

Many closely related species will not unite, or do so only with extreme difficulty, and distantly related species sometimes unite easily. Within one family there might be a genus – such as *Dianthus* – with species that cross readily, and another genus – such as *Silene* – in which the most persevering efforts to cross extremely close species have failed to produce even a single hybrid. This dichotomy exists even within the limits of one genus. For example, the many species of *Nicotiana* have been crossed more than the species of almost any other genus, but Gärtner finds that even though *Nicotiana acuminata* is not particularly distinct, it obstinately fails to fertilize or be fertilized by eight other *Nicotiana* species. There are many analogous examples.

No one has been able to establish how much or what kind of difference is sufficient to prevent two species from crossing. There are plants with very different habits and general appearances, with different flower parts, cotyledons, fruit, and even pollen, that *can* be crossed. Annual and perennial plants, deciduous and evergreen trees, and plants inhabiting different habitats and fitted for extremely different climates can often be crossed with ease.

A "reciprocal cross" between two species means, for example, a stallion horse first being crossed with a female ass, and then a mare horse being crossed with a male ass.[2] There is often a huge difference in the facility with which the two parts of a reciprocal cross can be made. Such cases are very important because they prove that the capacity of any two species to cross is often completely independent of systematic affinity or any recognizable similarities in their whole organization. These cases also show that the capacity for crossing is related to constitutional differences confined to the reproductive system and imperceptible to us. Kölreuter long ago observed the contrast between the two parts of a reciprocal cross. To give one example: *Mirabilis jalappa* can easily be fertilized by pollen from *Mirabilis longiflora*, and the resulting hybrids are sufficiently fertile, but Kölreuter tried more than two hundred times over eight years to fertilize *Mirabilis longiflora* with pollen from *Mirabilis*

2. [In other words, male of species X crossed with female of species Y and female of species X crossed with male of species Y. Therefore, a "reciprocal cross" is actually two crosses with both possible male/female combinations. – D.D.]

Jalappa and utterly failed. Thuret makes the same observation for certain seaweeds. Moreover, Gärtner finds that some difference between the two parts of a reciprocal cross is very common, even in forms so closely related that many botanists rank them as only varieties (e.g., *Matthiola annua* and *Matthiola glabra*). It is also remarkable that the two sets of hybrids raised from the two parts of a reciprocal cross generally differ in fertility – sometimes to a great degree. Yet the two crosses that produce them are between the same two species with only the positions of mother and father reversed.

Gärtner provides several other singular rules. For example, some species cross with others remarkably well, and other members of the same genus impress their characteristics on hybrid offspring remarkably well; however, these two powers do not necessarily go together. Some hybrids always closely resemble only one of their parent species rather than being intermediate between the two, as is usual. Although such hybrids are externally very much like one of their parents, with rare exceptions they also tend to be extremely sterile. Likewise, among hybrids that are usually intermediate in structure between the parents, exceptional and abnormal individuals sometimes closely resemble only one parent, and these are almost always completely sterile (even when the other hybrids raised from seed from the same capsule are fertile). These observations demonstrate that the degree of fertility in the hybrid is completely independent of its external resemblance to either pure parent.

These rules governing fertility in first crosses and of hybrids show that when distinct species are united, their fertility graduates from zero to perfect, and even, under certain conditions, to excess. Besides being susceptible to favorable and unfavorable conditions, their fertility is innately variable. The degree of fertility is not necessarily the same in the first cross and the hybrids that result. Hybrid fertility is not related to how much the hybrids resemble either parent. Lastly, the ease with which a first cross between two species can be made is not always governed by their systematic affinity or resemblance. This final statement is clearly proven by reciprocal crosses, because there is generally some difference, and occasionally a great difference, in how easily a union can be effected depending on which species is used as the mother and which the father. The hybrids resulting from reciprocal crosses also differ in fertility.

Do these complex and singular rules indicate that species have been endowed with sterility simply to prevent them from confusedly mixing in nature? I think not. For why should sterility between various pairs of species differ so much?

Wouldn't it be equally important to keep *all* species from blending together? Why should the degree of sterility vary innately among individuals of the same species? Why should some species cross easily but produce very sterile hybrids while other species cross with difficulty but produce fairly fertile hybrids? Why should the results of the two parts of a reciprocal cross often differ so much? We can even ask, why are hybrids permitted at all? To grant species the special ability to produce hybrids, and then to stop their propagation by varying degrees of sterility, not strictly related to how easily the first cross occurred, seems a strange arrangement.

In fact this discussion clearly indicates that sterility in first crosses and of hybrids is simply incidental to or dependent upon unknown differences, chiefly in the reproductive systems, between the species being crossed. These differences are so peculiar and limited that in reciprocal crosses between two species, the male sexual element of one species will often function on the female sexual element of another species with the reverse arrangement being ineffectual.

That sterility is incidental to other differences and is not a specially endowed quality can be explained more fully through an example. Whether or not a plant can be grafted onto another is totally unimportant to its welfare in the wild, so I am sure no one would argue that this capacity is specially endowed, but rather that it is incidental to the rules governing the growth of the two plants. Sometimes the reason one tree will not take on another is evident from differences in their rates of growth, the hardness of their wood, the period of the flow or nature of their sap, or other factors, but in many cases there is no apparent reason. A major difference in the size of two plants, one being woody and the other herbaceous, one being evergreen and the other deciduous, or in their adaptation to very different climates does not always prevent the two forms grafting together. As with hybridization, the capacity for grafting is limited by systematic affinity: no one has been able to graft together trees belonging to different families, and closely related species and varieties of the same species can usually, but not always, be easily grafted. However, as with hybridization, the capacity for grafting is not absolutely governed by systematic affinity. Although members of many distinct genera belonging to the same family have been grafted together, sometimes species of the same genus will not take on each other. The pear can be grafted far more easily on the quince (which belongs to a different genus) than on the apple (which belongs to the same genus). Even different varieties of pear take differently on the quince, and so do different apricot and peach varieties on certain varieties of the plum.

Just as Gärtner found that different individuals of the same two species sometimes have innate differences in crossing, so Sagaret believes that different individuals of the same two species differ in how readily they can be grafted together. There is also a parallel to reciprocal crossing: how easily a graft can be made depends on which of the two species is donating the graft. For example, the common gooseberry cannot be grafted on the currant, whereas the currant will take on the gooseberry, albeit with difficulty.

The sterility of hybrids, with their imperfect reproductive organs, is very different from the difficulty of uniting two pure species, which have perfectly functional reproductive organs. Yet these distinct cases run parallel to a certain extent. Something analogous occurs in grafting: Thouin finds that three species of *Robinia* that seed freely on their own are rendered barren after being grafted onto another species. Yet the grafting itself presents no difficulty. Meanwhile, certain species of *Sorbus* yield twice as much fruit than usual when grafted onto other species. This second example recalls the extraordinary case of *Hippeastrum*, *Lobelia*, and other plants that seed more freely when fertilized by pollen from distinct species than when self-fertilized by their own pollen.

Despite the obvious and fundamental difference between mere adhesion of grafts and the union of male and female sexual elements during reproduction, there is a rough parallel in the results between grafting and crossing. Just as the complex rules governing how easily trees can be grafted onto each other are incidental to unknown differences in their vegetative systems, so the even more complex rules governing how easily first crosses can be made are incidental to unknown differences, chiefly in the reproductive systems. In both cases these differences roughly follow systematic affinity – which, after all, attempts to capture every kind of similarity and difference among organisms. The facts do not indicate that the ease or difficulty of crossing or grafting various species is a special endowment, yet the difficulties of crossing are as important for the endurance of specific forms as the ease of grafting is unimportant for their welfare.

The difficulty of a first cross apparently depends on several distinct factors. Sometimes it is physically impossible for the male sexual element to reach the ovule, such as when a plant has a pistil that is so long that the pollen tubes cannot reach the ovarium. Or when pollen from one species is placed on the stigma of a distantly related species, the pollen tubes protrude but do not penetrate the stigmatic surface.[3] Even if the male sexual element reaches the female sexual element, it may be incapable of initiating the formation of an embryo; this seems to be the case with some of Thuret's experiments on seaweeds. These observations cannot be explained any more than why certain trees cannot be grafted onto others. Lastly, the embryo might develop but perish at an early stage. This possibility has not been sufficiently explored. It is a common cause of species sterility based on observations by Mr. Hewitt, who has a lot of experience hybridizing gallinaceous birds. At first I did not believe this, because once hybrids are born they are generally healthy and long-lived – just think of the common mule. However, hybrids face different circumstances before and after birth. If they are born into a region where both parent species live, they face a suitable environment. A hybrid possesses only half the qualities of its mother's species, so it faces partly unsuitable conditions while it is nourished in the womb, egg, or seed provided by the mother and thus may be liable to perish at an early stage. Moreover, all very young organisms seem particularly susceptible to unnatural environments.

The case of hybrid sterility, where sexual elements are imperfectly developed, is very different. I have already alluded to the large set of observations I have collected showing that when plants and animals are

3. [The female portion of a flower includes the pistil, which is made up of a pollen-catching tip called the stigma and a stalk leading to the ovaries. When pollen, which carries sperm, sticks to the stigma, it grows a "pollen tube" to transport the sperm to an ovule within an ovary. – D.D.]

removed from their natural environment their reproductive systems are liable to become seriously affected. In fact, this is the great bar to the domestication of animals. There are many similarities between this kind of sterility and hybrid sterility. In both cases sterility is independent of general health and is often accompanied by excess size or great luxuriance. In both cases the sterility occurs to various degrees and it is most likely to affect the male element (although the reverse also occurs), and it tends to go along with systematic affinity. Entire groups of plants and animals can be rendered impotent by the same unnatural conditions, and there are entire groups of species that tend to produce sterile hybrids. Sometimes a single species in a group survives major environmental changes with its fertility unimpaired, and certain species within a given group produce unusually fertile hybrids. It is impossible to tell without trying if a particular animal will breed in confinement, a plant will produce seed in culture, or if any two species in a genus will produce more or less sterile hybrids. Lastly, organisms placed in an unnatural environment for several generations become variable; I believe this is because their reproductive systems are especially affected (though not as severely as when sterility ensues). The same is true of hybrids: they tend to vary with successive generations, as every experimentalist has observed.

So when an organism is placed into a new and unnatural environment, and when hybrids are produced by the unnatural crossing of two species, the reproductive system is similarly affected by sterility – independently of general health. In the first case the environment is disturbed, though often so slightly as to escape notice. In the second case (that of hybrids) the environment remains the same, but the organization is disturbed by two different structures and constitutions being blended together. Indeed, it hardly seems possible that two organizations could be compounded into one without some disturbance to development, or to the relationships between the organs, or to the relationship of the organs to the environment. When hybrids are able to breed with each other, they pass the same compounded organization to offspring, so it is not surprising that sterility rarely diminishes with successive generations, even if it is slightly variable.

There are only vague hypotheses to explain several puzzling observations about hybrid sterility. Examples include the unequal fertility of hybrids produced by the two parts of a reciprocal cross, and the greater sterility of those occasional hybrids that closely resemble one of the two parent species. I do not pretend that my remarks in this chapter get to the root of the matter. I do not try to explain why an organism becomes sterile when it is placed in unnatural conditions. All I have tried to show is that sterility is the common result in two somewhat related cases: when an organism's environment is disturbed, and when an organism's organization is disturbed by two forms of organization having been compounded into one.

It may seem fanciful, but I suspect that a similar parallelism extends to a related but different set of observations. There is an old and almost universal belief founded on a considerable body of evidence that slight changes to the environment are beneficial to all living things. Farmers and gardeners act on this in their frequent changes of seeds, tubers, and so forth, from one soil or climate to another and back again. Convalescent animals benefit greatly from any changes in habit. There is abundant evidence for both plants and animals that a cross between very distinct individuals of the same species (i.e., between different strains or subbreeds) endows the offspring with vigor and fertility. Indeed, as I mentioned in chapter 4, a certain amount of crossing is indispensable, even for hermaphroditic species, whereas inbreeding for several generations between closely related individuals, especially if they are kept in identical environments, always induces weakness and sterility in the progeny.

It seems therefore that slight changes to the environment benefit all organisms, and crosses between slightly different individuals of the same species give vigor and fertility to offspring; but major changes, or particular types of changes, often render organisms sterile to some degree, and crosses between very different individuals or different species generally produce sterile hybrids. I cannot convince myself that this parallelism is an accident or an illusion; this series of observations seems connected by some common but unknown bond that is fundamentally related to the principle of life.

A powerful argument can be made that there must be some essential distinction between species and varieties and that there must be some error pervading the above remarks, because varieties cross without difficulty and yield fertile offspring regardless of how much they may differ in external appearances. I admit that this is almost always the case. But if we consider varieties produced in the wild, we are immediately entangled in hopeless difficulties. If two forms previously thought to be varieties are found to be at all sterile when crossed, most naturalists immediately rank them instead as separate species. The blue and red pimpernel, the primrose, and the cowslip are considered varieties by many botanists, but Gärtner finds that they are not quite fertile when crossed, so he ranks them as species. If we argue thus in a circle, then all varieties in the wild are fertile by definition.

Turning to varieties produced or supposed to have been created under domestication does not clarify the issue. The German Spitz dog crosses with foxes more easily than other dogs do, and certain dogs indigenous to South America do not cross readily with European dogs; the explanation that occurs to everyone is that these dogs have descended from multiple original species (this is probably correct). It is nevertheless remarkable that there are so many groups of domestic varieties that appear very different but are perfectly fertile together (e.g., varieties of the pigeon or the cabbage). This is especially striking given how many species closely resemble one another and yet are completely sterile when crossed.

The fertility of domestic varieties is actually less remarkable than it appears at first. For one thing, the external dissimilarity of two species does not determine their degree of sterility when crossed, and the same rule applies to domestic varieties. Second, some eminent naturalists maintain that a long course of domestication eliminates sterility with each successive generation of hybrids, which were at first only slightly sterile. If this is true, then sterility cannot both appear and disappear under nearly the same conditions. Lastly, the most important consideration is that under domestication, new plant and animal varieties are generated through the methodical and unconscious power of human selection. Breeders do not and could not select for slight differences in the reproductive system or other constitutional differences correlated with the reproductive system. They provide varieties with the same food, treat them nearly the same way, and do not seek to alter their habits. Nature acts slowly and uniformly over vast periods of time on the whole organization in any way that is for each creature's good; nature may thus modify the reproductive system in the descendants of each species, either directly, or more probably indirectly, through correlation. With this distinction between the process of selection by humans and by nature, the different results should not be surprising.

I have so far implied that varieties of the same species are invariably fertile when crossed. But it is impossible to resist the evidence for some sterility in the following cases.[4] For several years Gärtner kept a type of dwarf maize with yellow seeds and a tall variety with red seeds growing next to each other in his garden. Although these plants have separate sexes, they never crossed by themselves. He then fertilized thirteen flowers of the one with pollen from the other, but only a single head produced any seed – only five grains. Manipulation could not have been harmful in this case, because the plants have separate sexes. No one considers these varieties of maize to be distinct species, and because the hybrid plants raised by the cross were perfectly fertile, even Gärtner did not venture to consider the two varieties as two species.

Girou de Buzareingues crossed three varieties of a gourd, which has separate sexes like the maize. He asserts that the greater the difference between two varieties, the greater the difficulty of mutual fertilization. I do not know how far these experiments can be trusted, but Saraget ranks these three gourd types as varieties, and he bases his classifications mainly on the test of fertility.

The following case is amazing. It is based on an astonishing number of experiments made during many years on nine species of *Verbascum*. The experiments were carried out by so good an observer and so hostile a witness as Gärtner. He found that when yellow and white varieties of the same species of *Verbascum* are crossed, they produce less seed than when either colored variety is fertilized with pollen from its own flowers. Furthermore, he asserts that when yellow and white varieties of one species are crossed with yellow and white varieties of a *different* species,

4. It is at least as persuasive as the evidence used to establish the sterility between many species, and derived from hostile witnesses, who in all other cases consider fertility and sterility as safe criteria for distinguishing between species.

more seed is produced by crosses between the same colored flowers than by crosses between differently colored flowers. Yet the only distinction between these varieties is flower color, and one variety can sometimes be raised from the other's seed. Based on my own observations, I suspect that certain hollyhock varieties present an analogous case.

Kölreuter, whose accuracy has been confirmed by every subsequent observer, has proven the remarkable fact that one variety of the common tobacco is more fertile when crossed with a widely distinct species than are other varieties. He conducted experiments with five forms of tobacco that are usually ranked as varieties. He tested them in the most rigorous way – with reciprocal crosses – and found that their hybrid offspring are perfectly fertile. But when one of these five varieties was used as either mother or father in a cross with *Nicotiana glutinosa* (a very different species), it always yielded hybrids that were more fertile than those produced from crosses between the other four varieties and *Nicotiana glutinosa*. Thus, the reproductive system of this one variety must have been modified somehow.

So it is difficult to ascertain the fertility between varieties in the wild, because if a supposed variety is sterile at all, its rank is automatically changed to "species." Furthermore, human selection is based on external characteristics without the goal of producing undetectable functional differences in the reproductive system. Based on these considerations, I conclude that the fertility of varieties is not universal, does not form the fundamental distinction between varieties and species, and does not overthrow the view I have put forward that the general sterility in first crosses and of hybrids is incidental to slowly acquired modifications, especially to the reproductive systems of the crossed forms. Thus, sterility is not a special endowment.

The offspring of crossed species (here referred to as "hybrids") and the offspring of crossed varieties ("mongrels") can be compared independently of the question of fertility. Gärtner sought to draw a very distinct line between species and varieties, and he could find only few and unimportant differences between the so-called hybrid offspring of species and the so-called mongrel offspring of varieties; indeed, they are similar in many important ways.

The most important distinction is that in the first generation, mongrels are more variable than hybrids. However, Gärtner admits that hybrids of species that have been cultivated for a long time are often variable in the first generation, and I have seen striking cases of this myself. Gärtner further admits that hybrids springing from closely related species are more variable than those from very distinct species, showing that the difference in variability graduates away. Propagation of mongrels and of particularly fertile hybrids for several generations produces extremely variable offspring, but in a few cases both hybrids and mongrels long maintain a uniformity in their characteristics. Variability through successive generations is perhaps greater in mongrels than hybrids, which is not surprising. The parents of mongrels are varieties – usually domestic varieties, because very few experiments have been conducted on natural varieties – implying that in most cases there has been some recent variability. It is not surprising that such variability would often continue and that it would be added to the variability arising from crossing.

The hybrids that spring from a first cross are less variable than hybrids in successive generations. This observation is curious and deserves attention. It corroborates my view on the cause of ordinary variability – namely, that it results from the sensitivity of the reproductive system to environmental changes. These often cause impotence or at least an inability to properly produce offspring identical to the parent form. A first-generation hybrid is the offspring of two species that have normal, unaffected reproductive organs (except in the case of species that have been cultivated for a long time), but the hybrids themselves have seriously affected reproductive organs and their descendants are highly variable.

However, to return to the comparison of mongrels and hybrids: Gärtner states that mongrels are more likely to revert to either parent form than hybrids are. But if this is true, it is only a difference in degree. Gärtner further insists that when any two species, even when closely related, are crossed with a third species, the two resulting hybrids are very different; whereas when two distinct varieties of one species are crossed with another species, the two resulting hybrids are not very different. As far as I can make out, however, this conclusion is based on a single experiment and contradicts the results of several experiments by Kölreuter.

These are the only differences – and they are trivial – that Gärtner can point out between hybrid and mongrel plants. Moreover, according to Gärtner, the rules that govern the resemblance between hybrids and their parents – especially if the parents are closely related species – are the same as those that govern the resemblance between mongrels and their parents. When two species are crossed, one of them is sometimes predominant in impressing its likeness on the hybrid offspring; the same is true when plant varieties are crossed. Hybrid plants produced by a reciprocal cross generally resemble each other, as do the mongrels produced by a reciprocal cross. Both hybrids and mongrels can be reduced to either pure parent species by repeated crosses with a parent species over successive generations.

These concepts are apparently applicable to animals, but this subject is very complicated partly because of secondary sex characteristics and mostly because the ability to transmit likeness to offspring predominates in one sex. This is true both when one species is crossed with another species and when one variety is crossed with another variety. For example, the ass has a predominant power over the horse, so both the mule and the hinny resemble the ass more than the horse. (The prepotency is stronger in the male ass than in the female, because a mule is more like an ass than is the hinny.)

Some authors stress that mongrel animals, but not hybrid animals, closely resemble one of their parents. But this does in fact occur with hybrids, though much less frequently than with mongrels. I have collected cases of crossbred animals that closely resemble one parent, and the resemblances seem chiefly confined to monstrous characteristics that appear suddenly, such as albinism, melanism, missing tail or horns, or extra fingers and toes. These are not characteristics slowly acquired by selection. Consequently, sudden reversions to the perfect characteristics of either parent are more likely to occur with mongrels, which are often the offspring of suddenly produced semi-monstrous varieties, than with hybrids, which are the offspring of slowly descended and naturally produced species. I agree with Dr. Prosper Lucas, who arranged an enormous body of observations concerning animals and concludes that the rules governing the resemblance of offspring to parents are the same whether the parents differ greatly from each other (two species) or only slightly (two varieties).

Laying aside the question of fertility and sterility, there is a general similarity between the offspring of crossed species and the offspring of crossed varieties. This similarity would be astonishing if we supposed that species were specially created and varieties were the result of secondary laws, but it harmonizes perfectly with the view that there is no essential distinction between species and varieties.

In summary, the first crosses between individuals from distinct species are often sterile; when they are not, the resulting hybrid offspring are. The sterility is of all degrees, though it is often so slight that the two most careful experimentalists who have ever lived have come to opposite conclusions in ranking organisms this way. Sterility varies innately with each individual and is susceptible to favorable and unfavorable conditions. The degree of sterility does not strictly follow systematic affinity and is governed by curious and complex rules. Sterility is generally different – sometimes very different – between the two parts of a reciprocal cross, and it is not necessarily the same in a first cross and in the resulting hybrid.

In the grafting of trees, the capacity of one species or variety to take on another is incidental to generally unknown differences in their vegetative systems; similarly in crossing, the ease with which one species unites with another is incidental to unknown differences in their reproductive systems. Arguing that species have been endowed with various degrees of sterility to prevent blending in nature is like arguing that trees have been specially endowed with analogous difficulties in grafting to prevent them from growing into one another in forests.

Sterility in first crosses between pure species, which have perfectly functional reproductive systems, depends on several circumstances and in some cases mostly on early death of the embryo. Sterility of hybrids, which have imperfect reproductive systems because their entire organization is a compound of two distinct species, seems closely related to the sterility that affects pure species when their environment is disturbed. This is supported by another parallel: the crossing of forms that are slightly different favors the vigor and fertility of their offspring, and slight environmental changes are favorable to the vigor and fertility of all organisms. It is not surprising that the ease of making a first cross, the fertility of the resulting hybrids, and the capacity for grafting all run mostly parallel to systematic affinity, for this attempts to capture all kinds of resemblances among species.

First crosses between forms known to be varieties, and their resulting mongrel offspring, are generally fertile – also not surprising given how varieties in the wild are defined, and given that most varieties produced under domestication arise by the selection of external characteristics and not characteristics of the reproductive system. Excluding the question of fertility, hybrids and mongrels are closely comparable in all other respects. The observations given in this chapter do not contradict and even support the view that there is no fundamental distinction between species and varieties.

9

THE IMPERFECTION OF THE
GEOLOGICAL RECORD

IN CHAPTER 6 I LISTED THE MAIN OBJECTIONS TO THE IDEAS
presented in this book. Most of them have now been discussed. One – the
distinctness of species and the fact that they do not blend together via
countless transitional links – is an obvious difficulty. I have given reasons as to why such links cannot be observed in places that would seem
to favor their existence (i.e., large continuous areas with a graduated
environment). I tried to show that the survival of each species depends
more on the presence of already-defined species than on climate, and
this significant aspect of the environment does not graduate away like
heat or moisture. I also tried to show that intermediate varieties exist in
smaller numbers than the forms they connect and are therefore generally
beaten out and exterminated during further modification and improvement. The main cause for the lack of many existing intermediate links
is the very process of natural selection, through which new varieties
continually replace their parent forms. However, this extermination acts
on a massive scale, so the number of intermediate varieties that has ever
existed must be enormous. Why, then, isn't every geological formation
and stratum full of extinct intermediate links? Geology does not reveal
a finely graded chain of organisms, and this is perhaps the gravest objection to my theory. The explanation lies in the extreme imperfection of
the geological record.

First, bear in mind the kind of intermediate forms that my theory
predicts must have existed. When considering two species I find it difficult not to imagine forms *directly* intermediate between them. But this
is entirely false; we should always look for forms that are intermediate

between each species and a common but unknown ancestor.[1] Moreover, the common ancestor probably differed in some ways from all of
its modified descendants. Consider a simple illustration: the fantail and
pouter pigeons both descended from the rock pigeon. If we possessed
all the intermediate varieties that ever existed, we would have a close
series between the rock pigeon and the fantail and another close series
between the rock pigeon and the pouter. *But there would be no varieties
directly intermediate between the fantail and the pouter.* For example, none
would exhibit a combination of both expanded tail and enlarged crop,
the distinguishing features of these two breeds. Furthermore, these two
breeds have become so modified that if there were no historical or indirect evidence as to their origins, it would be impossible to determine
by simply comparing their structures with the rock pigeon's that they
actually descended from it and not from some other species.

Likewise, if we consider two very distinct forms in the wild, like the
horse and the tapir, there is no reason to suppose that there were ever
any direct links between them. Rather, the forms we would seek would
be between each and some unknown common ancestor. The common
ancestor would have had some general resemblance to both the tapir and
the horse, but in some structures it may have differed considerably from
both, perhaps more than they differ from each other now. In all such
cases it would be impossible to recognize the ancestral form of any two

---

1. [This is an exceptionally important point. Refer to the inset accompanying
the tree diagram in chapter 4. It depicts a small evolutionary tree, where D is at 20
million years ago, C at 10 million years ago, and A and B are at the present. A and B are
two related species, and C is their common ancestor, which means that both A and B
descended from C. For much of the book, Darwin discusses linear evolution, or what he
calls "descent with modification." In the diagram, this is the process by which species
D becomes species C, and species C becomes species A, *or* the process by which D
becomes C, which becomes B. How C splits into two new linear paths leading to A or B
represents speciation, or "the origin of species." Darwin's point is that there are no direct
intermediates between A and B: notice in the diagram, there is nothing there! There *are*
intermediates between A and C and between B and C.

The diagram also illuminates another question discussed previously: how much
separation between the A and B lines is necessary before the two forms are distinct
species instead of varieties? Darwin argues that there really is no essential difference
between varieties and species. Varieties are simply incipient species, or forms on the way
to becoming species. – D.D.]

or more species – even if we could compare the ancestor's structure with that of its modified descendants – unless we possessed a nearly perfect chain of the intermediate links.

My theory does not necessarily prohibit the possibility of one living form having descended from another living form – for example, a horse from a tapir – and in such a case direct intermediates would have existed between them. However, this would imply that one of the forms had remained the same for a very long time while its descendants had changed greatly. Competition between organisms and competition between parent species and descendant species will make this rare, because new and improved forms tend to supplant old and unimproved forms.

According to the theory of natural selection, every living species of a genus is connected to the parent species of that genus by differences no greater than those between varieties of an extant species. These parent species, now generally extinct, are connected in turn to more ancient species, and so on back through time, always converging on the common ancestor of each great class. The number of intermediate and transitional links between all living and extinct species must therefore be inconceivably great, but if this theory is correct, then they have all assuredly lived upon this earth.

Besides the problem of not finding fossil remains of all these links, there is the objection that not enough time has passed to allow such a great amount of slowly effected organic change through natural selection. It is difficult for me to impress upon the reader the facts that can lead the mind to feebly comprehend the lapse of time. The person who can read Charles Lyell's *Principles of Geology* – which future historians will recognize as having produced a revolution in natural science[2] – but deny the incomprehensible vastness of past time may close this book at once. Not that it would be enough to study the *Principles of Geology,* or to read special treatises on different geological formations and note how each author attempts to give an idea of the duration of each formation or even each stratum. A man must spend years examining great piles of super-

2. [Darwin precisely predicts Lyell's legacy. Note that Darwin himself was a highly accomplished geologist. – D.D.]

imposed strata for himself and watch the sea grinding down old rocks to make fresh sediment before he can hope to comprehend anything of the lapse of time, whose monuments we see around us.

It is good to wander along coastlines formed of moderately hard rocks and contemplate the process of erosion. In most cases the tides reach the cliffs for only a short time twice a day, and the waves eat into the cliffs only if charged with sand or pebbles. (Pure water probably has little or no effect.) If the base of a cliff is eventually undermined, then huge fragments fall down and become fixed. These are then worn away atom by atom until they are so reduced in size that the waves can roll them around; then they can be ground more quickly into pebbles, sand, or mud. The bases of retreating cliffs are often littered by rounded boulders thickly clothed by marine organisms, showing how little they are abraded and how seldom they are rolled about. If a rocky cliff undergoing degradation is followed for many miles, the portions that are currently suffering are found only here and there, along a short length or around a promontory. The vegetation and appearance of the surface everywhere else show that years have passed since waters washed their base.

Anyone who closely studies the action of the sea on shorelines will be deeply impressed by how slowly rocky coasts are worn away.[3] With this impression in mind, consider the beds of conglomerate, many thousands of feet thick. These probably formed more quickly than many other deposits and yet contain worn and rounded pebbles, each bearing the stamp of time, thereby showing how slowly the mass had accumulated. Recall Lyell's profound remark that the thickness and extent of sedimentary formations measure a corresponding degradation elsewhere in the earth's crust. And what massive amounts of degradation sedimentary deposits of many regions imply! Professor Ramsay has given me the maximum thickness of each formation in Great Britain (usually from actual measurements, but in a few cases from estimates):

Paleozoic strata (not including igneous beds)	57,154 ft.
Secondary strata	13,190 ft.
Tertiary strata	2,240 ft.

---

3. The observations made by Hugh Miller and Mr. Smith of Jordan Hill are impressive.

These values add up to nearly 13.75 British miles! Some of these formations are represented by thin beds in England but are thousands of feet thick on the European continent. Moreover, according to most geologists, enormously long "blank" periods passed between each formation, so the lofty pile of sedimentary rocks in Britain provides but an insufficient perception of the time that passed during its accumulation – yet what time this must have consumed! Careful observers estimate that the great Mississippi River deposits sediment at a rate of only six hundred feet per hundred thousand years. This figure may not be correct, but given how extensively oceanic currents spread fine sediment, accumulation in any one area must be extremely slow.

The erosion suffered by strata in many places, independently of the rate of accumulation of the eroded material, probably offers the best evidence of the lapse of time. I was struck by its effects when I saw volcanic islands that the waves have pared all around into cliffs one or two thousand feet high. The gentle slope of the solidified lava streams reveals how far the hard, rocky beds had once extended into the open ocean. The same story is told even more clearly by faults, those great cracks along which the strata rise up on one side and plummet on the other to the height or depth of thousands of feet. But the land has been so completely planed down by the action of the sea since the surface cracked that no trace of these vast dislocations is externally visible. For example, the Craven Fault extends for more than thirty miles, and the vertical displacement of the strata varies from six hundred to three thousand feet along its length. Professor Ramsay has published an account of a downthrow in Anglesea of twenty-three hundred feet, and he informs me that there is one in Merionethshire of twelve thousand feet. Nothing on the surface reveals these displacements, because the piles of rock on either side have been swept away. Contemplating these observations stirs my mind almost as much as vainly grappling with the concept of eternity.

I am tempted to give one other case, the well-known erosion of the Weald, even though it is insignificant compared to the erosion that has removed masses of Paleozoic strata, in some places ten thousand feet thick, as shown in Professor Ramsay's masterly work on the subject. Yet it is interesting to stand on the North Downs and look at the distant South Downs; not too far to the west, the northern and southern escarpments meet and close, and one can easily imagine the great dome of rocks that must have covered the Weald as recently as the later part of the deposition of the Chalk formation. The distance from the North to the South Downs is about twenty-two miles, and Professor Ramsay informs me that the average thickness of the formations is about eleven hundred feet. Some geologists claim that a range of older rocks underlies the Weald, on the flanks of which the overlying sedimentary deposits might have accumulated more thinly than elsewhere. If this is true, then the above estimate of thickness is incorrect, but this source of doubt probably doesn't affect the western extremity of the region. If we knew the rate at which the sea usually wears away a line of cliff at any given height, we could measure the time needed to erode the Weald. Of course we lack this information, but in order to form some rough idea of the process, assume that the sea eats into a five-hundred-foot-high cliff, simultaneously along its whole length, at the rate of one inch per century. This may seem a small allowance at first, but it is the same as assuming that a one-yard-high cliff gets eaten back along a whole coast at a rate of one yard per twenty-two years. I doubt any rock, even one as soft as chalk, would yield so quickly. (Highly exposed coasts may be an exception, and the degradation of lofty cliffs is probably more rapid from the breakage of fallen fragments.) Furthermore, no line of coast of ten or twenty miles ever suffers degradation along its entire length simultaneously, and almost all strata contain harder layers that resist erosion and form a breakwater. Therefore, under normal circumstances, erosion of a five-hundred-foot-tall seaside cliff at one inch per century is a generous estimate, but even at this rate the erosion of the Weald must have taken 306,662,400 years – that is, about 300 million years!

The erosive action of freshwater on the once upraised, now gently inclined Wealden region could not have been great, but it would reduce this estimate. This area is known to have undergone oscillations in level, so the surface may have existed as dry land for millions of years, escaping the action of the sea. When deeply submerged, it would have escaped the coastal waves for equally long periods. In all probability, then, far more time has passed since the later part of the Secondary period than 300 million years.[4]

4. [Darwin overestimates here by about 100 million years, but his point is to illustrate the huge stretches of time involved. – D.D.]

I have commented on these examples because it is very important for the reader to gain some idea of time's great passage. During each of these many years, the land and the water across earth were inhabited by hosts of living things. How many generations must have succeeded one another in the long roll of years! It is a number incomprehensible, impossible for the mind to grasp. Now turn to the richest natural history museums, and behold the meager display!

Everyone knows our fossil collections are incomplete. The late paleontologist Edward Forbes remarked that the number of known fossil species is based on single and often broken specimens, or on a few specimens collected from one spot. Only a small portion of the earth's surface has been geologically explored, and no one area has been examined completely, as proven by the important discoveries made in Europe every year. An organism with an entirely soft body cannot be preserved; shells and bones decay and disappear at the bottom of the sea, where sediment does not accumulate. It is a mistake to assume that sediment accumulates uniformly over most of the ocean floor quickly enough to embed and preserve fossil remains. A bright blue tint across a large portion of the ocean bespeaks the purity of the water. There are many formations that were covered by others only after vast periods of time had elapsed, yet had not suffered wear and tear. This seems explicable only if the seabed often lies in an unaltered condition for ages. Remains that become embedded in sand or gravel are usually dissolved by rainwater when the beds are upraised. I suspect that very few of the many animals that live on the beach between the high and low watermarks are preserved. For example, several species of Chthamalinae – a subfamily of rock barnacle – coat rocks along shores all over the world. Their populations are huge. Only a single Mediterranean species inhabits deep water, and its fossils have been found in Sicily; but not one other species of the group has been found in any Tertiary formation even though the *Chthamalus* genus existed during the Chalk period. The mollusk genus *Chiton* presents a partially analogous case.

It is unnecessary to point out that the fossil evidence of terrestrial organisms from the Secondary and Paleozoic periods is extremely fragmentary. No land shells have been found from either of these vast periods, with one exception discovered by Sir Charles Lyell in the Carboniferous strata of North America. Likewise, a single glance at the historical table in the supplement to Lyell's *Manual of Elementary Geology* shows how accidental and rare the preservation of mammals is. Nor is this rarity surprising, given that a large proportion of the fossils of Tertiary mammals has been discovered in either lake-bed deposits or caves, but not a single true lake bed or cave is known dating from the age of Secondary or Paleozoic formations.

However, the imperfection of the geological record results mainly from the huge gaps of time between formations rather than anything discussed above. Reading about the formations, or seeing them in nature, makes them seem closely consecutive. But Sir R. Murchison's work in Russia reveals huge gaps between successive formations, and similar observations apply to North America and many other parts of the world. If a skillful geologist confined his studies to only one of these areas, he would never suspect that periods that were blank and barren in his own country had seen the accumulation of sediment charged with new and peculiar life forms elsewhere. The amount of elapsed time between each formation cannot be ascertained in any specific area.[5] The common and significant differences in the mineral composition of consecutive formations implies major changes in the geography of the surrounding land (the source of sediments) and agrees with the hypothesis that vast periods of time pass between each formation.

But I think it is easy to see why the geological formations of each region are almost invariably intermittent. While examining hundreds of miles of South American coast that had been upraised several hundred feet within the recent period, I was struck by the absence of any recent deposits sufficiently extensive to survive even a short geological period. Along the whole western coast, Tertiary beds are so scantily developed that no record of several successive and peculiar marine faunas seems likely to survive to a distant age. Muddy streams and large-scale erosion of coastal rocks supply copious sediments on South America's western shore, and yet this rising coast presents no extensive formation with recent or Tertiary fossils. The explanation must be that as coastal and

---

5. [Modern geological science has almost surmounted this difficulty, especially with the advent of atomic dating. – D.D.]

sub-coastal deposits are brought up by the gradually rising land, they are continually worn away by waves grinding against the shore.

Sediment must accumulate in extremely thick, solid, or extensive masses to withstand the incessant action of waves upon upheaval, as well as during the subsequent oscillations in level. Such thick and extensive deposits can form in two places: (1) the profound depths of the sea (where, according to the research of E. Forbes, there are few animals, so when a deposit like this becomes upraised, it will provide an extremely imperfect record of life forms that once lived there), or (2) in a shallow bottom, if it subsides continuously, allowing for thick deposits of sediment. In the second case, if the rate at which the bed subsides remains approximately equal to the rate at which sediment is supplied, the sea will remain shallow and favorable to life. This may result in a fossil-rich formation that is thick enough to resist degradation when upraised. I am convinced that all the ancient fossil-rich formations formed this way.[6]

All geological observations clearly reveal that each area has undergone numerous slow oscillations in level that apparently affected large regions. Consequently, during periods of subsidence, formations that are rich in fossils and sufficiently thick to withstand degradation can form over wide areas, but only where the supply of sediment keeps the seabed shallow and preserves the remains of organisms before they decay. But if the seabed remains stationary, *thick* deposits cannot accumulate along shallow parts, which are the most favorable to life. The formation of thick deposits is even less likely during periods of elevation, for the accumulated beds are destroyed when they are upraised and brought within the range of coastal action. Thus, the geological record is necessarily patchy. I am confident in these views because they agree with Charles Lyell's general principles of geology and because E. Forbes independently arrived at a similar conclusion.

6. I published my ideas on this subject in 1845. Since then, I have noticed that author after author has concluded that this or that great formation was accumulated during subsidence (i.e., formation mechanism 2). The only ancient Tertiary formation on the western coast of South America bulky enough to resist degradation up to now was definitely deposited during a downward shift in level. It gained its thickness this way, but it will hardly last to a distant geological age.

To continue, during periods of uplift, land area often increases, creating new habitats favorable to the formation of new varieties and species, as previously explained. But these periods are generally blank in
the geological record. During subsidence, land area and the number of
inhabitants decrease.[7] Extinction occurs and few new varieties or species form, yet it is exactly during these periods of subsidence that great
fossil-rich deposits are created. It is almost as if Nature guards against
the frequent discovery of her transitional forms.

These considerations show that the geological record as a whole
is extremely imperfect. But it is more difficult to understand why any
particular formation does not contain the remains of varieties linking
species that lived at its commencement and its close. (There are a few
cases of one species presenting distinct varieties in the upper and lower
parts of one formation, but these are exceptional.) Although each formation requires vast numbers of years for deposition, I can see several
reasons why each should not contain a graduated series of fossil links.
However, I am unable to assign due proportional weight to the following
considerations.

Each formation may mark a long lapse of time, but perhaps less than
that required for one species to change into another. The paleontologists Bronn and Woodward conclude that the average duration of each
formation represents a period two or three times longer than the average
duration of a species, but several issues prevent a concrete conclusion
based on this observation. Just because a species appears in the middle
of a formation does not mean it did not previously exist somewhere else.
Similarly, just because a species disappears before the uppermost layers
of a deposit does not mean it became completely extinct at that time. It is
easy to forget how small Europe is compared with the rest of the world;
moreover, the various layers of the same formations throughout Europe
have not yet been perfectly correlated.

Marine animals of all kinds migrate during climatic and other
changes. Thus when a marine species first appears in any formation, its
appearance probably indicates the time it first migrated into that area

7. Except species that inhabit coastal regions when the shores of a continent first
break up into an archipelago.

rather than the time when it first appeared as a species. For example, some species are recorded earlier in the Paleozoic beds of North America than in those of Europe, apparently because some time was required for their migration from American to European seas. In examinations of the latest deposits, geologists have everywhere noted that still-extant species may be common in a deposit but have become extinct in the immediately surrounding sea, or, conversely, that species currently abundant in the surrounding seas are rare in or absent from a particular deposit.

Reflect on the established amount of migration by European species, on the major changes in level, on the exceptionally great changes in climate, and on the vast lapse of time during just the glacial period, which represents only a portion of one whole geological period. Yet it is unlikely that sedimentary deposits, including fossil remains, accumulated continuously in any one area during the whole of this period. For example, sediment was probably not deposited throughout the entire glacial period near the mouth of the Mississippi within the depth limits at which marine animals can flourish, because major geographic changes occurred in other parts of America during that time. When the beds that were deposited in shallow water near the mouth of the Mississippi do become upraised, fossils will first appear and disappear at different levels in different beds because of migration and geographic changes. A geologist examining these beds in the distant future might be tempted to conclude that the average duration of species with embedded fossils had been less than that of the glacial period when in fact it had really been much longer (extending from before the glacial period until today).

In order to get a perfect gradation between a form in the lower layer of a formation and a related form in the upper layer of the same formation, a deposit must accumulate for a long enough period to allow variation and selection to occur: it must be thick. The species undergoing modification have to live in the same area throughout this whole time, which can only happen if the depth remains relatively constant. This in turn requires a continuous supply of sediment at a rate that counterbalances subsidence (recall that thick, fossil-rich formations develop only during periods of subsidence). However, subsidence tends to sink the region from which sediment is derived, diminishing supplies while the downward movement continues. In fact, a balance between the supply of sediment and subsidence is rare; several paleontologists have observed that very thick deposits are usually devoid of organic remains except near their upper or lower limits.

It seems that each individual formation has generally been accumulated intermittently, like the whole pile of formations in any one region. A formation made up of beds with differing mineralogical compositions indicates that the process of deposition was frequently interrupted; altered currents in the sea and a supply of sediment of a different nature generally result from long-term geographical changes. Even a very close inspection of a formation does not reveal how much time its deposition had consumed. There are many examples of beds that are only a few feet thick in one place but represent a formation that is thousands of feet thick elsewhere and must have taken an enormous amount of time to accumulate; yet no observer of just the thinner section would recognize the vast lapse of time it represents. There are many examples of the lower beds of a formation having been upraised, eroded, submerged, and then re-covered by the upper beds of the same formation; huge but easily overlooked intervals have interrupted their accumulation. In other cases great fossilized trees, still upright as they grew, provide evidence of the many long intervals of time during the process of deposition – intervals that would never have been suspected without them. Sir Charles Lyell and Mr. Dawson found fourteen-hundred-foot-thick Carboniferous beds in Nova Scotia with ancient root-bearing strata layered one above the other with at least sixty-eight different levels. This suggests that when the same species occurs at the bottom, middle, and top of a formation, it probably did not live on the same spot during the entire process of deposition. Instead, it disappeared and reappeared many times during a single geological period. A species that undergoes considerable modification during a geological period will therefore probably not leave a finely graded record of the intermediate forms that must have existed but will instead present abrupt, if slight, changes of form.

Recall that naturalists do not have a golden rule to distinguish species from varieties. They grant each species a small amount of variability, but if two forms are significantly different, they are ranked as species unless a series of close intermediates connects them. For reasons discussed above, this is rarely possible to discover in any one geological section.

Suppose species A is found in a bed below species B and C. Even if A is intermediate between B and C, it would be ranked as a distinct species unless some intermediate forms linked it to B and/or C. A might actually be the common ancestor of B and C without being completely intermediate between them in structure. So it is possible to find a parent species and its modified descendants all in one formation, not recognize their relationship, and rank them all as entirely separate species!

Many paleontologists found their species on exceptionally small differences, especially if the specimens come from different substrata of the same formation. Some experienced researchers of mollusk shells are now sinking the species defined by D'Orbigny and others to the rank of variety: evidence of the type of subtle change my theory predicts. Furthermore, fossils embedded in distinct but consecutive layers of one formation are typically ranked as species even though they are more closely related than species in more widely separated formations. I will return to this subject in the next chapter.

There is another consideration. Rapidly propagating plants and animals that do not move large distances generate varieties that generally remain local at first. Such local varieties do not spread widely nor replace their parent forms until considerable modification has occurred. Therefore, the chances of discovering all the early transitional stages between two forms contained within a formation are low, because successive changes will have taken place in some confined region. Most marine animals have a broad range, and those with the broadest are the most likely to spawn first local varieties and ultimately new species.[8] These ranges greatly exceed the sizes of known geological formations in Europe, so in this case, too, the chances of being able to trace transitional stages in one formation are low.

The current standard for linking two forms and declaring them one species requires many specimens of intermediate varieties from many places. Paleontologists can rarely accomplish this with fossil species. Consider a hypothetical example: Will future paleontologists be able to connect by fine, intermediate fossil links the lineages of different breeds

8. I have discussed that among plants, those with the broadest ranges most often generate varieties.

of cattle? Or sheep? Horses? Dogs? Will they have enough fossil evidence to deduce if these animals descended from single or multiple stocks? Will they be able to tell whether certain seashells on the shores of North America are species that are distinct from their European representatives, or just varieties? (Modern researchers cannot agree on the matter.) These future geologists will be able to grasp the answers only if many intermediate fossil gradations are found, and that is highly unlikely.

Geological research has added many species to existing and extinct genera. Although gaps between a few groups have been shrunk, the science has done little to break down the distinctions between species by connecting them with many intermediate varieties. This is the most obvious but also the most serious objection to my theory, so it is worthwhile to sum up the above discussion with a hypothetical illustration.

The Malay Archipelago is about the size of Europe (from Norway to the Mediterranean and Britain to Russia); it therefore equals all the closely examined geological formations except those of the United States. I fully agree with Mr. Godwin-Austen that the current state of the archipelago, with its many large islands separated by wide, shallow seas, is probably what Europe was like when most of its known formations were accumulating. The Malay Archipelago is one of the most biologically diverse regions on earth, yet if all the species that have ever lived there were collected together, they wouldn't even come close to representing the natural history of the world! Actually, the terrestrial organisms of the archipelago would probably be imperfectly preserved by the formations thought to be accumulating there. I suspect that few of the animals living in the littoral zone or those living on submerged rocks would be embedded, and those embedded in gravel or sand would not last to a distant epoch. No remains would be preserved wherever sediment did not accumulate on the seabed or where it did not accumulate quickly enough to stop the organisms from decaying.

Fossil-rich formations thick enough to last as far into the future as Secondary formations lie in the past could only be formed during subsidence. Such periods of subsidence would be separated from each other by long intervals of stasis or elevation. Fossil-rich formations would be destroyed during elevation by coastal erosion, as is now happening on the shores of South America. Subsidence generates fossils but also probably causes extinctions; elevation causes variation, but it also leaves an especially imperfect geological record. Even during favorable periods it is unlikely that subsidence and the accumulation of sediment that must go with it will continuously last as long as the average duration of a species. These contingencies are indispensable to the preservation of all the transitional gradations between any two or more related species. If these transitional forms are not preserved, fossils that actually *are* transitional will appear to be distinct species. Periods of subsidence are also likely to be interrupted by oscillations of level, and climatic changes would intervene during such long periods; the inhabitants of the archipelago would migrate, and no continuous record of their modifications would be preserved within one formation.

Many marine inhabitants of the archipelago currently range thousands of miles beyond its confines. I believe these far-flung species are the most likely to produce new varieties, which would initially be local but would eventually spread and replace their parent species if they were to become modified to any decided advantage. If one of these varieties were to return to its ancient home, it would be slightly different from its former state and would probably be ranked as a different species by paleontologists.

My theory predicts that all the past and present species of each group are connected into one long and branching chain of life, but we cannot expect to find infinite transitional forms in geological formations. We should reasonably expect to unearth only a few links; some closely related and some more distantly. I must admit I did not expect to learn that even the best-preserved geological sections offer such a poor record of the mutations of life.

The paleontologists Agassiz, Pictet, and especially Professor Sedgwick, have urged that the sudden appearance of whole groups of species in certain fossil-rich formations is fatal to the idea that species change. If many species belonging to the same genera or family really have at any point suddenly come into being, then the theory of descent with slow modification through natural selection must fall. The development of a group of related organisms descended from a single common ancestor must be a very slow process, and the common ancestor must exist long before its modified descendants. But the quality of the geological record is constantly overrated; just because certain genera or families have not been found below a certain stage does not mean they did not exist before that stage. It is easy to forget how large the world is in comparison to the area over which geological formations have been carefully examined. Groups of species may have existed and slowly multiplied in other places before invading the ancient archipelagoes of Europe and the United States. The enormous intervals of time *between* the deposition of each consecutive formation go underrated (even though some of these may be longer than the amount of time required for the accumulation of each formation). These intervals give record-free windows for the multiplication of species from one or a few parental forms, and in the succeeding formation all of these species would appear to have been suddenly created.[9]

A few examples will expose the error of supposing that whole groups of species appear suddenly. Works on geology published a few years ago cite the "well-known fact" that mammals appeared abruptly at the commencement of the Tertiary formations. But the richest known accumulation of fossilized mammals was recently discovered in the middle of a Secondary formation, and one mammal has been found in the new red sandstone at its start! Cuvier used to assert that no monkey fossils occurred in any Tertiary stratum, but extinct species have since been discovered in South America, India, and Europe, even as far back as the Eocene.

The whale family presents the most striking case. These animals have huge bones and range all over the world, but not a single whalebone had been discovered in any Secondary formation, which seemed to justify the notion that this great and distinct order suddenly came into being sometime between the deposition of the latest Secondary and earliest Tertiary formations. But the Supplement to Lyell's *Manual of Elementary Geology,* published in 1858, provides clear evidence of the

---

9. Recall that adaptation to a new and peculiar line of life – for example, flight in air – might require a long succession of ages, but once the change has been generated and a few species have acquired a great advantage over other organisms, a relatively short amount of time will see them produce many divergent forms, which would rapidly spread throughout the world.

existence of whales in the upper greensand deposited sometime *before* the close of the Secondary period.

I'll give another example, which particularly impressed me because it passed under my own eyes. In a work on fossilized rock barnacles, I discussed the number of extinct and extant Tertiary species, the extraordinary abundance of the individuals of many species from the Arctic to the equator in shallow and deep zones, the perfect way in which specimens from the oldest Tertiary beds have been preserved, and how easily even a fragment of a valve can be recognized. Based on these circumstances, I inferred that if rock barnacles had existed during the Secondary period, their fossils would definitely have been discovered. As not one species had been found in formations of that age, I concluded that this group developed suddenly at the start of the Tertiary. This troubled me as yet another example of a large group of species appearing abruptly. But my work had barely been published when the skillful paleontologist M. Bosquet sent me a drawing of an unmistakable rock barnacle fossil he extracted himself from the chalk of Belgium! As if to make the case as astonishing as possible, it was a *Chthamalus*, a very common and large genus of which no specimens had been found even in Tertiary strata. This example proves that rock barnacles existed during the Secondary period, and they may be the common ancestors to many Tertiary and existing species.

Fossilized teleostean fishes found low down in the Chalk period are the most commonly used example of a group of species apparently appearing suddenly. (This group includes the large majority of existing fish species.) Professor Pictet recently pushed their existence a substage back, and some paleontologists believe that certain much older fish with poorly understood relationships to other groups are actually teleostean. It would certainly be astonishing if the whole lot of them did in fact appear at the commencement of Chalk formation deposition, as Agassiz believes, but even this would not be insurmountable by my theory unless it could also be shown that the species of this group appeared suddenly and simultaneously throughout the entire world at this moment in geological time. Because of limited research there are very few specimens of fossilized fish from south of the equator. In fact, a run-through of Pictet's *Paleontology* reveals that very few species are known from several formations in Europe. Certain extant families of fish have a confined range; early teleostean fish might have been similarly confined before developing extensively in one area of the seas and then spreading. Furthermore, there is no reason to assume that the oceans were always as open from north to south as they are now. If the Malay Archipelago were converted into land, the tropical parts of the Indian Ocean would form a large enclosed basin in which any large group of marine animals may multiply. They would remain confined until some of the species became adapted to a cooler environment and could double the southern capes of Africa or Australia to reach distant seas.

Our ignorance of geology in regions outside Europe and the United States, the revolution in paleontological ideas brought about by recent discoveries, and the other considerations discussed above make dogmatization about the succession of organisms throughout the world about as rash as it would be for a naturalist who lands for five minutes in one barren place in Australia to discuss the number and range of the species of the entire continent.

A much graver difficulty is the sudden appearance of species, all belonging to the same group, in the lowest (i.e., deepest) known fossil-containing rocks. Most of the arguments that have convinced me that all the existing species within one group have descended from a common ancestor also apply to the earliest known species. For example, I am sure that all Silurian trilobites descended from some single crustacean that must have lived long before the Silurian and probably differed greatly from any known animal. Some of the most ancient Silurian animals, like the *Nautilus, Lingula,* and others, do not differ much from living species. But according to my theory, these old species cannot have been the common ancestors of all the species in the orders to which they belong, because they do not have characteristics that are in any way intermediate between them. Also, if they were the ancestors of these orders, then their many improved descendants would probably have replaced and exterminated them long ago.

So if my theory is correct, long periods of time must have elapsed before the lowest Silurian stratum was deposited. These vast periods were probably longer than the whole interval from the Silurian to the present, and they witnessed a world swarming with living creatures.[10]

I have no satisfactory explanation as to why records of this primordial age have not been found. Several eminent geologists, led by Sir R. Murchison, are convinced that the organic remains of the lowest Silurian stratum mark the dawn of life on this planet. Other judges, including Lyell and the late E. Forbes, dispute this conclusion. But only a small portion of the world is accurately known; M. Barrande recently discovered a lower stage to the Silurian system abounding with previously unknown and peculiar species. Traces of life have been detected in the Longmynd beds beneath Barrande's so-called primordial zone. Phosphate and bitumen have been found in some of the lowest azoic rocks, probably indicating that life existed when they were deposited.[11]

Making sense of the huge missing piles of fossil-rich strata that must have accumulated somewhere before the Silurian is a major difficulty. If these ancient beds have been eroded or otherwise obliterated by geological processes, only small remnants should remain, and these would generally be in an altered state. However, descriptions of Silurian deposits over immense regions of Russia and North America suggest that older formations do not necessarily suffer more from erosion and alteration. So the case is currently inexplicable and is a valid argument against the concepts entertained in this book, but to show that it might eventually receive some explanation, I give the following hypothesis.

Fossils from formations in Europe and the United States do not appear to have inhabited great depths, and the amount of sediment in these formations is miles thick. These two observations suggest that large islands or tracts of land existed in the neighborhood of the current

10. [Life is now known to have existed for billions of years before the periods discussed here; for most of that time all life forms were bacteria-like. Animal fossils date back around 200 million years before the time Darwin is referring to. – D.D.]

11. [Living things incorporate a lot of phosphorus (phosphates are phosphorus-containing salts), and bitumen is a hydrocarbon-rich goo that usually forms from large masses of ancient decomposed plants. "Azoic," wonderfully, means "without life." Paleontologists still use this category, but it now describes a much lower and more ancient set of strata than it did in the mid-1800s. – D.D.]

European and North American continents as a source of sediment. The geological state of these regions between the periods when the formations were deposited remains unknown: Europe and North America may have been dry land, submarine surfaces near land, or beds of an open and unfathomable sea.

Existing oceans are three times as extensive as the continents and studded with many islands, but not one of these islands has yet afforded even a remnant of Paleozoic or Secondary formations. This suggests that during the Paleozoic and Secondary periods, neither continents nor continental islands existed where the oceans now extend. Had they existed in these regions, then sediment derived from their surfaces would have accumulated into formations that would have been at least partially upheaved by oscillations of level. Oceans have extended where they extend today since the remotest period of which we have records, and large tracts of land have similarly existed in the Silurian, no doubt subjected to great oscillations of level, where continents exist now. The oceans are still mainly areas of subsidence, the continents of elevation, and the great archipelagos of oscillations. Has this been the case since eternity? Continents seem to form from the overall dominance of elevation during oscillations, but the dominant force may have changed in the lapse of ages. At a time immeasurably antecedent to the Silurian, continents may have existed where oceans now spread, and open seas may have existed where continents now stand. If the bed of the Pacific Ocean were now converted into a continent, we would not necessarily find formations older than Silurian strata. Such layers would have subsided some miles deeper and been pressed on by an enormous mass of water, possibly causing more severe changes to the rock than experienced by strata nearer to the surface. In some parts of the world, like South America, there are immense regions of bare metamorphic rock that must have experienced great heat and pressure, inviting some special explanation. These rocks may in fact predate the Silurian, in completely altered form.

I have discussed several serious challenges to my theory: infinitely numerous transitional links between all living and extinct species have not been found in successive formations, whole groups of species appear suddenly in European formations, and there are currently no known fossil-containing formations beneath Silurian strata. The gravity of these objections is illustrated by the unanimous and often vehement support of eminent paleontologists (Cuvier, Owen, Agassiz, Barrande, Falconer, E. Forbes, and others) and great geologists (Lyell, Murchison, Sedgwick, and others) for the immutability of species. (I have reason to believe that upon reflection, at least one – Sir Charles Lyell – entertains grave doubts about the accepted dogma.) I feel how rash it is to disagree with these great authorities, to whom, among others, we owe all our knowledge. Those who think the geological record is perfect and attach little weight to the observations and arguments I present in this book will reject my theory. For my part, following Lyell's metaphor, I consider the geological record an imperfectly kept history of the world written in an ever changing dialect. Of this entire history we possess only the last volume, relating to only two or three regions. In this one volume, only a few chapters have survived, and of each page, only a few lines here and there. Each word of the slowly changing language is different in the interrupted succession of chapters; these words represent the apparently abruptly altered forms of life entombed in consecutive but widely separated formations. In this reading, the challenges discussed above are diminished, or even disappear.

10

THE SUCCESSION OF ORGANISMS IN THE GEOLOGICAL RECORD

NEW SPECIES HAVE APPEARED VERY SLOWLY, ONE AFTER AN-
other, both on the land and in the waters. Sir Charles Lyell shows that
evidence for this in Tertiary stages is compelling, and every year tends
to fill up the blanks. In some of the most recent beds, only one or two
species have since gone extinct, and only one or two are new species
having appeared for the first time in that specific location, or as far as we
know, on earth. ("Recent" is a relative term. They would seem very old if
described in actual years.) If Philippi's observations in Sicily are correct,
the successive changes in the marine inhabitants there have been many
and most gradual. Secondary formations are more fragmented, but as
Bronn remarks, neither the appearance nor disappearance of their now-
extinct species is simultaneous in each formation.

Species in different genera or classes have not changed at the same
rate or to the same extent. The oldest Tertiary beds sometimes yield
shells of mollusks that still exist, surrounded by the shells of extinct
species. Falconer gives a similar case from the sub-Himalayan deposits;
he found the fossil of an extant crocodile species surrounded by many
strange and now lost mammals and reptiles. The Silurian *Lingula* differs
only slightly from living species in the same genus, while most of the
other Silurian mollusks and all the crustaceans have changed greatly.
Land dwellers seem to change more quickly than those of the sea – a
recent find in Switzerland provides a striking example – because organisms high on the scale of nature may change more quickly than those low
on the scale, although there are exceptions. As Pictet remarks, the rate
at which life on earth changes does not correspond to the succession of

geological formations; between any two successive formations, the various life forms rarely change in exactly the same degree. However, upon
comparison of all but the most closely related formations, all species are
found to have changed. And finally, once a species disappears from the
face of the earth, it never reappears. All of these observations agree with
my theory.

I do not believe that all species of a region can collectively change
abruptly, simultaneously, or to the same extent. Not only is the process of
modification slow, but also the amount of variability in one species is independent of the amount of variability in others. Whether or not natural
selection acts on this or that variation, and the extent to which variations
are accumulated, depends on many complex factors: on variations being
beneficial, on the power of crossing, on the breeding rate, on the slowly
changing physical environment, and on competition with other species.
This means that one species might not change, or change less, relative to
other species. We also find this when considering geographical distribution. For example, the land snails and coleopterous insects of Madeira
have come to differ significantly from their nearest relatives on the continent of Europe, whereas the birds and marine snails have remained
unaltered. I mentioned in a previous chapter that terrestrial and highly
organized organisms apparently change at a quicker rate than marine
and lower organisms because their relationships to the environment
are more complex. When many of the organisms in a region become
modified, competition and the all-important relationships with other
organisms mean that any form that does not become somewhat modified
and improved faces extinction. Consequently, all species within a region
do eventually become modified – if we consider long enough intervals.

During long and equal periods of time, the average amount of
change undergone by species belonging to one class may perhaps be
nearly the same. However, for reasons already discussed, the fossil record
is accumulated at wide and irregular intervals, so the fossils embedded
by consecutive formations present apparently unequal changes. Each
formation does not represent a new and complete act of creation, but
only an occasional scene, randomly taken, in a slowly changing drama.

It is understandable why a species never reappears once it has gone
extinct, even if the very same conditions recur. One species can become

adapted to fill the niche of another species and replace it, but the two forms cannot be identical, because each inherits unique characteristics bestowed by distinct ancestors. For example, imagine that all fantail pigeons were destroyed. By striving for long ages, pigeon breeders could probably recreate the same breed from the wild rock pigeon. But if the rock pigeon – the ancestor of pigeon breeds – also became extinct, then the fantail could not be made from any other pigeon species. Any fantail-like breed that might be produced would inherit characteristics from its ancestor that the fantail never had.

Entire genera and families follow the same rules as individual species in appearing and disappearing. A group does not reappear once it has disappeared, and its existence is continuous for as long as it lasts; this is also a prediction of my theory. (There are a few apparent exceptions, but so few that even E. Forbes, Pictet, and Woodward, opponents of my views, agree to these rules about groups.) Because all the species of a group have descended from a common ancestor, it is clear that for as long as any species of that group exist, the group must also continuously exist in order to generate either new and modified forms or the same old unmodified forms. On this basis, the genus *Lingula* must have existed continuously from the early Silurian to the present.

A genus or family tends to grow gradually, generating more species, until the group reaches a maximum. Eventually the number of species decreases as they become extinct. If the number of species in a genus or the number of genera in a family is represented by a vertical line of varying thickness, crossing successive geological formations in which the species are found, the line would gradually thicken upward, sometimes maintaining its thickness, until it thinned out in the upper beds. Sometimes the bottom of the line would falsely *appear* to begin abruptly at some thickness rather than as a sharp point. The process is necessarily gradual, because modification and the creation of related forms are slow. First, a species must spawn two or three varieties, which are slowly converted into species, which in turn gradually produce other species, and so on, like the branching of a great tree from a single stem.

Until now, I have only discussed the extinction of species and groups of species in an incidental way. Yet by the theory of natural selection, the production of new forms is intimately connected to the disappearance of old forms. The old notion that all the inhabitants of the earth have been swept away at successive periods by catastrophes has generally been abandoned, even by geologists whose general views would lead them to it (Elie de Beaumont, Murchison, Barrande, and others). Studies of Tertiary formations reveal that, instead, species and groups of species disappear gradually, one after another, first from one spot, then from another, and finally from the world. Both species and groups of species exist for varying amounts of time, with some groups having endured from the earliest known dawn of life to the present, whereas others disappeared before the close of the Paleozoic. No fixed law governs the length of time that a species or group will persist. The complete extinction of a species or group is generally slower than its production, although in some cases extermination has been wonderfully sudden, as with the ammonites toward the close of the Secondary period.

The whole subject of extinction has been mired in gratuitous mystery. Some authors have even suggested that each species has a definite duration just as an individual organism has a definite life span. I don't think anyone has marveled at the extinction of species more than I have. In La Plata I found the tooth of a horse embedded with the remains of *Mastodon, Megatherium, Toxodon,* and other extinct monsters that all coexisted with shelled mollusks that are still around. I was astonished. Horses have run wild and proliferated greatly since the Spanish introduced them to South America, so what could have exterminated the former horse in such an apparently favorable environment? Professor Owen recognized that although the tooth looks like that of a modern horse, it actually belonged to an extinct species. If this horse were still living but rare, no naturalist would be surprised by its rarity, an attribute common to many species in all classes and all regions. If a species is rare, there must be something unfavorable in its environment; as to what that thing is, we can hardly ever tell. Continuing with the supposition that this horse still exists but is rare, analogy with other mammals and the naturalization of the domestic horse in South America suggest that under more favorable conditions it would have quickly spread over the whole continent. It would not be possible to tell what conditions would have checked its increase, whether one or multiple factors would have contributed, nor at what period in the horse's life they would have acted and to what degree. We would not notice the conditions slowly becoming less favorable. The horse would have become rarer and rarer, and finally extinct, its niche seized by some more successful competitor.

It is easy to forget that the increase of every species is constantly checked by unperceived and adverse agencies, which are also sufficient to cause rarity and extinction. Many cases from Tertiary formations show that rarity precedes extinction, and the same progression is known for species driven to local or global extinction by humans. I will repeat something I published in 1845: to appreciate that species generally become rare before going extinct while at the same time marveling greatly when a rare species actually does go extinct is like appreciating that sickness precedes death but at the same time suspecting that when a sick man dies, he met his end by some unknown deed of violence.

The theory of natural selection is grounded on the belief that each new variety, and ultimately each new species, is produced and maintained because of some advantage over competitors. The consequent extinction of forms that are less favored follows almost inevitably. It is the same with domesticated organisms. When a new and slightly improved variety is raised, it first supplants other varieties in the immediate neighborhood; if it improves further, it is transported further, like short-horn cattle, and replaces other breeds. Thus, the appearance of new forms and the disappearance of old forms are bound together. In some flourishing groups the rate at which new species are generated may exceed the rate at which old ones are exterminated. But the total number of species has not been increasing indefinitely, at least during the later geological periods, so on long time scales it seems likely that the generation of new forms has caused the extinction of about the same number of old forms.

New species are most likely to exterminate their parent species, because competition is most severe between similar organisms. If many new forms develop from one species, the most closely related species – those in the same genus – are most likely to face termination. I believe this is how a new genus comes to supplant an old genus of the same family. However, a species from one group *can* seize the niche occupied by a species from a *distinct* group and thereby cause its extermination; if additional forms develop from the successful intruder, then many will have to yield their places. The suffering forms will likely be related, having inherited common inferiorities; a few of these may escape severe competition, and thus persist, if they are fitted to some peculiar niche or inhabit an isolated habitat. For example, a single species of *Trigonia*, a genus that left behind many shells in Secondary formations, survives in Australian seas, and a few members of the great and almost extinct ganoid fishes still inhabit fresh waters. So the total extinction of a group is generally slower than its formation.

Concerning the apparently sudden extermination of whole families or orders, like the trilobites at the close of the Paleozoic and the ammonites at the close of the Secondary, recall that consecutive formations are punctuated by large intervals of time when there might have been extensive but slow extinction. Furthermore, when a new group of species immigrates to an area suddenly or develops unusually rapidly, it will exterminate many of the old inhabitants correspondingly suddenly. The yielding forms will commonly be related, partaking of some inferiority in common.

So the way in which single species and groups of species become extinct accords well with the theory of natural selection. Extinction should not dumbfound us. But what *should* is our presumption of imagining for even a moment that we understand the complex contingencies on which the existence of each species depends. The entire system of nature is based on the tendency of each species to increase, always checked yet seldom perceived by us. When we understand exactly why one species is more abundant than another, why certain species can be naturalized in certain regions while others cannot, and still be unable to account for the extinction of particular species and groups of species, *then* we will have the right to feel dumbfounded!

The geological discovery that life forms change almost simultaneously throughout the world is astonishing. Deposits that correspond to the European Chalk formation are found in parts of the world where there is not a fragment of the mineral chalk itself. Fossils from these beds in North America, equatorial South America, Tierra del Fuego, the Cape of Good Hope, and India unmistakably resemble fossils from the Chalk beds. In some cases not a single species is identical, but they all belong to the same families, genera, and sections of genera. Sometimes the similarities are superficial, like shape. Moreover, fossil forms not found in European Chalk beds but occurring in formations above and below are also absent in the deposits that correspond to the Chalk. Several authors have noticed a similar parallelism in the Paleozoic formations of Russia, Western Europe, and North America. Lyell states the same about Tertiary deposits in Europe and North America. Even if the few fossil species common to the Old and New Worlds are ignored,[1] the striking similarities among analogous groups of fossils would still be obvious and the formations could easily be correlated. These observations, however, are restricted to marine organisms. There isn't sufficient data to judge if land and freshwater organisms also change simultaneously across the world.[2]

When I write that marine life forms change "simultaneously," I do not mean at the same hundred-thousandth or even thousandth of a year. I do not even mean it in a strict geological sense. Imagine that all the marine animals currently living in Europe and all those that lived in Europe during the Pleistocene (an enormously remote period that includes the whole glacial period) were compared with those *currently* living in South America or Australia; the most skillful naturalist would be unable to decide whether it was the Pleistocene or the modern inhabitants of Europe that resembled those of the southern hemisphere more closely. Similarly, several able observers believe that the existing organisms of the United States are more closely related to those that lived in Europe during certain Tertiary periods than to those living in Europe now. This means that fossil-rich beds currently being deposited on North American shores may in the future be classified with older European beds. But all of the more modern marine formations – namely, the upper Pliocene, the Pleistocene, and the latest beds of Europe, North America, South America, and Australia – would correctly be ranked as "simultaneous" by some geologist of the distant future; they would all contain related

---

1. ["New World" refers to the Americas, whereas "Old World" refers to Europe, Asia, and Africa. – D.D.]

2. In fact, this type of parallelism can be doubted, although correlations can still be made. If the *Megatherium, Mylodon, Macrauchenia,* and *Toxodon* were transplanted to Europe from La Plata without any information about which formations contained their fossils back in South America, no one could tell that they had coexisted with still-extant shelled mollusks. But because these anomalous monsters coexisted with the *Mastodon* and horse, it might at least have been inferred that they lived during one of the later Tertiary stages.

fossils and would not contain forms found only in the older underlying deposits.

Mr. Edouard de Verneuil and Mr. Adolphe d'Archiac have also been struck by the fact that life forms change simultaneously – in the large sense I explained above – at distant parts of the world. After referring to the parallelism of Paleozoic life forms in various parts of Europe, they add, "If struck by this strange sequence, we turn our attention to North America, and there discover a series of analogous phenomena, it will appear certain that all these modifications of species, their extinction, and the introduction of new ones, cannot be owing to mere changes in marine currents or other causes more or less local and temporary, but depend on general laws which govern the whole animal kingdom." (Barrande has made forcible remarks to the same effect.) It is indeed futile to try to understand these great mutations in the forms of life throughout the world by looking to changing currents, climate, or other physical conditions; as Barrande has remarked, we must look to some special law. (This will be clearer when I consider the present distribution of organisms and find only a tenuous relationship between the physical conditions of a region and the nature of its inhabitants.)

The theory of natural selection explains this parallel succession of life forms throughout the world. New species arise from new varieties that possess some advantage, and forms already dominant are the most likely to generate new varieties (i.e., incipient species). There is evidence for this among plants, as those that are dominant – common in their native region and widely diffused – generate the greatest number of new varieties. It is also natural that dominant, varying, and far-spreading species should be the ones with the best chance of spreading further and giving rise to varieties and species in new regions. Diffusion may often be slow because it depends on climatic and other physical changes, but in the long run, dominant forms generally succeed in spreading. Land animals on distinct continents probably spread more slowly than marine organisms in the continuous sea, which explains the less strict parallel succession of terrestrial fossils.

A dominant and spreading species may encounter an even more dominant species, and then its triumphant course, or even existence, would cease. The precise conditions most favorable to the multiplication

of a new and dominant species are unknown, but a few factors are apparent. A large number of individuals increases the chances that favorable
variations will appear. Severe competition with many already existing
forms is also favorable, as is the ability to spread into new regions. Recurrent isolation at long intervals is probably also favorable, as explained
before. If two large regions provide equally amenable circumstances,
then whenever their inhabitants meet the resulting battle will be long
and intense. Eventually the most dominant forms prevail, regardless
of where they originated. Their victory causes the extinction of inferior forms, and because these inferior forms are related by inheritance,
whole groups tend to slowly disappear. (Though here and there a single
member might sometimes persist for a long time.) And so the principle
of dominant species spreading and spawning new species while causing
extinction is consistent with the parallel and simultaneous succession of
the same life forms throughout the world.

There is one more worthwhile consideration on this topic. I have
already discussed why large fossil-rich formations were deposited during
subsidence, that long periods during which the seabed was stationary
or rising have left blanks in geological records, and that fossils have not
been preserved where sediment was not accumulated quickly enough to
embed and preserve organic remains. During these long blank intervals,
I assume that organisms still experienced considerable modification and
extinction, and there was much migration. Because large regions are
affected by the same geological movements, strictly contemporaneous
formations have often been accumulated over wide spaces in the same
part of the world. But there is no reason to assume that large areas have
invariably been affected by the same movements; if two formations were
deposited in two regions during nearly, but not exactly, the same period,
then there should be essentially the same succession of life forms in both,
as discussed. However, the species would not correspond exactly, because there would have been slightly more time in one region for modifications, extinctions, and migrations.

There are probable cases of this phenomenon in Europe's geological history. In his memoirs Mr. Prestwich draws a close general parallel
between successive Eocene deposits in England and France. However,
when he compares certain stages between the two countries, he finds a

curious accordance in the numbers of species from the same genera, but the species themselves differ in a way that is difficult to explain given the proximity of the two regions (unless one assumes that an isthmus separated two seas inhabited by distinct but contemporaneous faunas). Lyell makes similar observations on some of the later Tertiary formations. Barrande shows that there is a striking parallel between successive Silurian deposits of Bohemia and Scandinavia, but he nevertheless finds a surprising amount of difference in the species. One explanation is that formations in these regions were not deposited during exactly the same periods. A layer in one formation may correspond to a blank in another, and if species change *continuously*, then formations in two regions could be arranged in the same order and the order would appear to be perfectly parallel, even though not all the species would be the same in the apparently corresponding stages.

All extinct and living species belong to one grand natural system, and this fact is directly explained by the principle of common descent. As a general rule, the more ancient an organism, the more it differs from living organisms. However, as Buckland remarked long ago, all fossils can be classified into still-extant groups or as intermediate between them. Extinct forms of life clearly plug the wide gaps between existing genera, families, and orders. The series is much less perfect if confined to just the living, or just the extinct, than if both are combined into one general system. With respect to the vertebrates, whole pages could be filled with striking illustrations by the great paleontologist Richard Owen showing how extinct animals fall between existing groups. Cuvier ranked ruminants and pachyderms as the two most distinct mammalian orders, but Owen has discovered so many fossil links that he has had to alter their whole classification, placing some pachyderms in the same suborder as ruminants. For example, he dissolves with fine gradations the apparently wide difference between the pig and the camel. Barrande – and there is no higher authority – asserts that although Paleozoic invertebrates belong to the same orders, families, or genera as existing animals, they were not limited in such distinct groups as they are now.

Some writers object to extinct species or groups of species being considered intermediate between living species or groups of species.

The objection is probably valid if "intermediate" is taken to mean that an extinct form can be *directly* in between two living forms in all characteristics. But I think it is clear that in a perfectly natural classification, many fossil species must stand between living species, and some extinct genera must stand between living genera (even between genera from different families). The most common case, especially with respect to very distinct groups such as fish and reptiles, seems to be that two groups currently distinguished by many characteristics were distinguished in the distant past by somewhat fewer.

It is a common belief that the more ancient a form, the more its characteristics tend to connect groups now widely separated, but this view should be restricted to groups that have undergone extensive change over geological ages. In any case, it would be difficult to prove the proposition, because even living animals are occasionally discovered with affinities to very distinct groups (for example, the lungfish). Comparisons between the older reptiles, amphibians, fish, cephalopods, and Eocene mammals and their more recent counterparts reveal that there is nonetheless some truth to the idea.

I'll refer back to the tree diagram of chapter 4 to illustrate how these observations and inferences accord with the theory of descent with modification. Let the numbered letters represent genera, and let the lines diverging from them represent the species in each genus. (The diagram is too simplistic – too few genera and too few species – but this is not important for illustrative purposes.) Let the horizontal lines represent successive geological formations, and let all the forms below the top line be extinct. The three existing genera, a^{14}, q^{14}, and p^{14}, form a small family; b^{14} and f^{14} form a closely related family or subfamily; and o^{14}, e^{14}, and m^{14} form a third family. These three families, together with the many extinct genera on the several lines of descent from the common ancestor A, form an order – all will have inherited something in common from their common ancestor. The principle of divergence during common descent (which I previously used this diagram to illustrate), predicts that the more recent a form is, the more it tends to differ from its ancient ancestor. This explains why the most ancient fossils differ the most from existing organisms. But do not assume that divergence is a necessary contingency; it depends solely on the descendants of a species being enabled to seize many different niches. A species may be modified only slightly in relation to slightly altered conditions and yet retain the same general characteristics through a vast period, as illustrated by several Silurian forms. On the diagram it is represented by the example of F^{14}.

The diagram can be used to depict many extinct forms embedded in successive formations. If fossils were discovered low down in the series, then the three existing families on the top line would seem less distinct. For example, if the genera a^1, a^5, a^{10}, f^8, m^3, m^6, and m^9 were discovered in fossil form, the three living families would be so closely linked that they would probably have to be united into one large family (as with Owen's ruminants and pachyderms). Yet the objection to calling the extinct genera "intermediate" is justified, because they are not directly intermediate, but by a long circuitous course through many different forms. If the fossils of many extinct forms were discovered above one of the middle geological formations – say, above the horizontal line labeled 6 – but not below it, then only the two families on the left would have to be united while the other two families would remain distinct. (However, these two families would be less distinct than before the discovery of the fossils.) This example built on the diagram illuminates why extinct genera are often somewhat intermediate between their modified descendants or between their collateral relations.

In reality the situation is far more complicated than the diagram suggests; groups are more numerous, endure for very unequal lengths of time, and are modified to various degrees. We possess only the last volume of the geological record, and it's very incomplete. Thus, except in rare cases, we should not expect on this scant information to fill wide gaps in nature's system and unite distinct families or orders. We can only expect that groups that have undergone major modifications in known geological periods should be more similar to one another in the older formations. Evidence from the best paleontologists corroborates this expectation.

The theory of descent with modification explains the main observations concerning the affinities of living and extinct forms. They are wholly inexplicable on any other view.

This theory also predicts that fauna from any major period in earth's history will be intermediate between their predecessors and successors.

In the diagram, species that lived during the sixth stage are the modified descendants of those that lived during the fifth stage and the ancestors of those in the seventh stage, so they can hardly fail to be intermediate between the forms above and below. But allowance must be made for extinction of some forms, appearance of new ones by migration, and a large amount of modification during the long blank periods in between successive formations. One example is enough: when paleontologists first discovered fossils of the Devonian, they immediately recognized them as intermediate between those of the overlying Carboniferous and underlying Silurian formations. But the fauna are not necessarily exactly intermediate, because the intervals of time that passed between the depositions of various formations were not equal. Anyone acquainted with the current distribution of species across the globe would not suggest that intermediacy is due to static physical environments. Recall that oceanic life forms have changed almost simultaneously throughout the world, in a vast range of physical environments. Also consider that climatic vicissitudes of the Pleistocene period (which encompassed the whole glacial period) affected species of the sea only slightly.

The theory of descent gives an obvious meaning to the fact that fossil remains from consecutive formations are often closely related. As I tried to show in the last chapter, the accumulation of deposits is often interrupted, leaving blanks in the record between geological formations. Preservation of all the intermediates between fossil species found at the top and bottom of a formation is therefore unlikely. However, we should unearth closely related forms, or "representative species," separated by intervals that are very long if measured in years but represent only a short time geologically. This is what we do find, and it is evidence for the slow mutation of species.

Some genera are exceptions to the general rule of intermediacy. For example, Dr. Falconer arranged *Mastodons* and elephants in two series, first according to mutual affinities and then according to their periods of existence. The two arrangements do not agree, as the species with the most extreme characteristics are not the oldest or most recent, and those with intermediate characteristics are not intermediate in age. To compare great things to small, if all living and extinct domestic pigeon varieties were put in order based on characteristics, the arrangement would not be the same as their order of production or disappearance. Even if the recorded appearance and disappearance of species in some geological formation agrees exactly with their actual development and demise, species do not necessarily endure for equal lengths of time. A very ancient form may last longer than some relatively novel form (this is particularly likely of terrestrial organisms inhabiting different regions).

Closely connected to the concept of intermediacy is the observation of all paleontologists that fossils from two consecutive formations are more closely related than fossils from two remote formations. Pictet uses the well-known example of remains from the stages of the Chalk formation, which generally resemble each other but are from distinct species. This single fact seems to have shaken Professor Pictet's firm belief in the immutability of species!

Although naturalists have not yet agreed on the distinction between "high" and "low" organisms, there has been much discussion about whether or not more recent species are higher than ancient forms. In one particular sense, my theory suggests that more recent species must be higher than older species, because each new species is formed by possessing some advantage over preceding forms in the struggle for life. If the Eocene inhabitants of some part of the world were put into competition with existing species in that same region under a similar climate, the modern organisms would exterminate the Eocene fauna or flora. The process of improvement must have affected the organization of the more recent and victorious life forms, but I can see no way to test this sort of progress. Crustaceans, for example, are not the highest in their own class, but they may have beaten the highest mollusks. European species have recently spread over New Zealand and seized previously occupied niches, so it seems likely that if all the plants and animals of Great Britain were set free in New Zealand, many British species would become naturalized and exterminate a multitude of native forms. Conversely, very few organisms from the southern hemisphere have become wild in any part of Europe, so if all the plants and animals of New Zealand were set free in Great Britain, they would probably not seize a significant number of niches now occupied by native organisms. In this scheme the organisms of Great Britain are "higher" than those of New Zealand, yet even the most skillful naturalist examining species from the two countries could not have foreseen this result.

Agassiz asserts that, to a certain extent, ancient animals resemble embryos of recent animals in the same class, or that the geological succession of organisms is somewhat parallel to embryological development of recent forms. I agree with Pictet and Huxley that this hypothesis is very far from proven. I nevertheless expect to see it eventually confirmed, at least with subordinate groups that have branched off from one another relatively recently, because it accords well with the theory of natural selection.[3] In chapter 13 I will attempt to demonstrate that an adult differs from its embryonic form due to variations intervening at a late point in life, which, when inherited, appear at a corresponding age in offspring. This process leaves the embryo unaltered but, over the course of successive generations, adds more and more difference to the adult.

The embryo may be a sort of picture preserved by nature, recording the ancient and less modified condition of each animal, but this may not be something that can ever be proven. The oldest known mammals, reptiles, and fish belonging to their own proper classes are only slightly less distinct from one another than typical members of the same groups today. The search for animals with the common embryological characteristics of vertebrates is in vain until beds far beneath the lowest Silurian strata are discovered, an unlikely eventuality.

Many years ago Mr. Clift showed that fossil mammals from Australian caves are closely related to living Australian marsupials. Even to the untrained eye, a similar relationship is manifest in La Plata, South America, where gigantic pieces of armor have been found that are similar to those of the armadillo. Professor Owen has shown that most of

---

3. [These ideas cause perennial controversy and confusion because the correct observation that related organisms exhibit similar developmental pathways is often confounded with the discredited idea of "ontology recapitulating phylogeny." Embryonic development does *not* retell the evolution of a species. For example, a human embryo does not go through "fish," "reptilian," and "mammalian" stages, but its developmental program is similar to that of other vertebrates. Vertebrates stem from a common ancestor and therefore share many fundamental characteristics, including the nature of their developmental pathways. – D.D.]

the fossil mammals in La Plata, buried there in such numbers, are related to living South American organisms. This relationship is depicted
even more clearly by the wonderful fossil collection of Mr. Lund and
Mr. Clausen from the caves of Brazil. These observations prompted me
to assert in 1839 and 1845 a "law of the succession of types" and "this
wonderful relationship in the same continent between the dead and the
living." Professor Owen subsequently extended the same generaliza-
tions to mammals of the Old World, and the same law is manifest in his
restorations of the gigantic extinct birds of New Zealand and in the birds
found in the caves of Brazil. Mr. Woodward has shown the same rule to
hold with seashells, although this case is not as clear, because most mollusk genera are widely distributed. And there are other cases, including
the relationship between extinct and living land snails on Madeira, and
between extinct brackish-water mollusks of the ancient Aral-Caspian
Sea and their modern counterparts.

It would not make sense to compare the modern inhabitants of Australia and South America at the same latitude and ascribe their dissimilarities to differing physical environments while simultaneously claim-
ing that the uniformity of types during the later Tertiary results from a
similarity of conditions. Also consider that marsupials have developed
in places other than Australia, and Edentata and other American types
have developed outside South America. Numerous marsupials inhabited
ancient Europe, and I discuss in the publications mentioned above that
the distribution of terrestrial animals in America was different from
what it is now. North America used to have much more the character of
present South America. Similarly, Falconer and Cautley's discoveries
demonstrate that Indian mammals were once more closely related to
African mammals than they are now. There are analogous facts pertain-
ing to the distribution of marine animals.

In light of these observations, what is the meaning of the succession of the same types in a confined area? The theory of descent with
modification explains this rule. The inhabitants of some given region will
obviously tend to leave closely related, although somewhat modified descendants in that same region. If the inhabitants of one continent differ
greatly from another, then their descendants will also differ greatly, and
in nearly the same way. But after very long intervals, great geographical

changes, and inter-migrations, feeble forms yield to dominant forms, and the rules governing past and present distribution become plastic.

A potential question could be asked in ridicule: "Do you suppose the *Megatherium* and other related monsters left as degenerate descendants the sloth, armadillo, and anteater of South America?" No. These huge animals went extinct and left no progeny. But in the caves of Brazil lie fossils of many extinct species that are similar in size and other charac- teristics to living South American species. Some of these fossils may be the actual ancestors of existing animals. Recall that, according to my theory, all species of a genus have descended from one ancestral species. If six genera consisting of eight species each are found in a geological for- mation, and the succeeding formation contains six other related genera, it means that only one species of each of the six older genera left modified descendants, which came to constitute the six new genera. The other seven species of the old genera all died out and left no progeny. Or – and this probably occurs more often – two or three species of two or three of the older genera left modified descendants, thereby parenting six new genera with the other species and other whole genera going extinct. A failing order with decreasing numbers of species and genera – appar- ently the case with the Edentata of South America – will leave still fewer descendants.

In this and the preceding chapter I attempted to show that the geologi- cal record is imperfect; that only a small portion of the world has been geologically explored with care; that only certain classes of organisms have been largely preserved in a fossil state; that museum collections of fossil specimens and the species they represent are insignificant when compared with the incalculable number of generations that have passed during even the deposition of even a single formation; and that the inter- vals of time between the deposition of successive formations are enor- mous. There has probably been more extinction during periods of subsid- ence, and more variation during periods of elevation, when the record is poorly kept. Individual formations are not deposited continuously, and the duration of deposition is perhaps less than the average duration of a species. Migration plays an important role in the first appearance of new forms in any given area; widely ranging species vary the most and give rise to new species most often; and varieties are frequently local at first.

All of these causes taken together render the geological record extremely imperfect and explain why paleontologists do not find countless varieties connecting together all extinct and living life forms by fine gradations. I also mentioned the lack of fossil formations predating the first Silurian beds. I can only address this issue hypothetically: the oceans and oscillating continents have probably been in their present locations back to the Silurian period, but long before that the globe may have been very different. Continents composed of formations older than any known to us now may have been fundamentally altered or lie buried under the ocean. Anyone who rejects these views of the geological record rightly rejects my theory.

All the other major aspects of paleontology agree, it seems to me, with the theory of descent with modification through natural selection. The theory explains how new species appear slowly and successively and why species of different classes do not necessarily change together, at the same rate, or to the same degree. It also predicts that in the long term all organisms undergo some modification. The extinction of old forms follows inevitably as new species are generated, and once a species disappears, it never reappears. Groups of species increase slowly and endure for unequal periods of time because modification occurs slowly and depends on many complex contingencies. The dominant species of large dominant groups tend to leave many modified descendants, giving rise to new groups and subgroups. Because of their inferiority, inherited in common, species of less vigorous groups tend to become extinct together and leave no modified descendants. The total extinction of an entire group may nevertheless often be a very slow process, as a few descendants may linger in protected and isolated habitats. Once a group disappears entirely, it cannot reappear, because the chain of successive generations has been broken.

The theory explains how dominant life forms – those that most often vary – tend to spread modified descendants over the world. They generally succeed in supplanting species that are their inferiors in the struggle for existence. This is why after long intervals of time the earth's inhabitants will appear to have changed simultaneously.

All life forms, ancient and recent, make one grand system. All are linked by successive generations. Divergence of character explains why the more ancient a form, the more it generally differs from those now living. The theory explains why extinct forms apparently fill gaps between existing forms, sometimes blending two groups previously classed as distinct, but more commonly just bringing them a little closer together. (The more ancient a form, the closer it is to the common ancestor of now-distinct groups; ancient forms therefore exhibit characteristics that are apparently intermediate between these groups. Extinct forms are rarely directly intermediate between living organisms.) Fossils from consecutive formations are more closely related to each other than are fossils from remote formations, because they are more closely linked by successive generations. It is also clear why remains from an intermediate formation have intermediate characteristics.

The inhabitants of each successive period in the earth's history have beaten their predecessors in the race for life and are in this way "higher" on the scale of nature, which may account for that ill-defined sentiment among paleontologists that on the whole, organization has "progressed." This view may be rendered more intelligible if it can be proven that ancient animals resemble embryos of living organisms in the same class. The preservation of the same types of organisms within each area during later geological periods is no longer mysterious; it is simply explained by inheritance.

If the geological record is as imperfect as I believe it to be, and it definitely cannot be much better, then the main objections to the theory of natural selection diminish or disappear. All the main laws of paleontology proclaim that species have formed by ordinary generation: old forms replaced by new life, produced by the variation still present all around us, and preserved by natural selection.

11

THE GEOGRAPHICAL
DISTRIBUTION OF LIFE

THE SIMILARITIES AND DIFFERENCES AMONG ORGANISMS across the earth cannot be explained by climate or other physical conditions. Almost every author who has recently studied the subject comes to this conclusion. The American continent alone would suffice to prove the point. Excluding the polar regions, there is a fundamental division in geographical distribution between the New World and the Old World. Despite this division, the vast American continent includes very diverse conditions: humid regions, arid deserts, lofty mountains, grassy plains, forests, marshes, lakes, and great rivers, all across wide temperature ranges. There is hardly an environment in the Old World that has no parallel in the New (at least as closely as the same species would require).[1] The conditions of the Old and New Worlds are parallel – and yet the organisms are wildly different.

In the southern hemisphere, parts of Australia, South Africa, and western South America between the 25th and 35th southern parallels share similar environments, yet their inhabitants could not be more different. South American organisms south of the 35th parallel and those north of the 25th necessarily face considerably different climates, yet they are more closely related to one another than to the organisms of Africa or Australia under nearly the same climates. (Analogous observations could be listed for ocean dwellers.)

---

1. A group of organisms is rarely confined to a small spot with unique conditions. For example, there are small areas in the Old World that are hotter than any in the New World but are not inhabited by peculiar flora or fauna.

A second major and striking observation is that barriers to free migration have a bearing on the differences between the inhabitants of various regions. This is illustrated by the dramatic contrast between nearly all the terrestrial organisms of the eastern and western hemispheres. (The northern parts are exceptional in that the land is almost continuous and in a slightly different climate may have allowed for the free migration of northern temperate forms, like arctic forms today.) Likewise, there are huge differences between inhabitants of equal latitudes in Australia, compared with Africa, and with South America, and these continents are as isolated from one another as possible. The pattern is the same within each continent: lofty and continuous mountain ranges, large deserts, and sometimes even large rivers demarcate differences between inhabitants. However, these barriers are not as impassable or likely to have endured as long as the oceans partitioning the continents, so the resulting differences are not as prodigious.

The same rule applies to the seas. Hardly a shelled mollusk, fish, or crab is common to both the eastern and western shores of Central and South America, and yet these are separated only by the narrow but impenetrable Isthmus of Panama. West of America's shores, a wide stretch of open ocean is not punctuated by a single island as a halting place for migrating animals, and beyond this barrier, on the eastern islands of the Pacific, lives a totally distinct fauna. So here three marine faunas range far north and south in parallel swaths not far from each other and under corresponding climates, but they are separated by impenetrable land or sea barriers and consequently remain wholly distinct. But proceeding westward from the eastern islands of the tropical Pacific, there are no impassable barriers, and many islands can serve as stepping-stones, so between the eastern islands of the tropical Pacific and Africa's eastern shore there are no well-defined and distinct marine faunas. The three faunas roughly defined as Eastern American, Western American, and Eastern Pacific Island have very few shelled mollusks, crabs, or fish, in common, whereas many fish range from the Pacific Ocean to the Indian Ocean, and many shelled mollusks are common to the eastern Pacific islands and Africa's eastern shore (on almost opposite lines of longitude).

A third major observation, implied by previous statements, is that organisms in the same sea or on the same continent are related to one another, even though individual species from different places and habitats are distinct. This is a very general rule and every continent offers countless examples. A naturalist traveling from, say, north to south will always notice the striking succession of specifically distinct organisms that are nevertheless clearly related. He will hear from closely related but unique birds notes that are nearly the same; spy similarly constructed nests, although these nests will not be quite alike; and observe eggs that are almost, but not quite, the same color.

The plains near the Straits of Magellan are inhabited by one species of rhea (American ostrich). Another species of rhea lives on the plains of La Plata to the north – a latitude where, in Africa, you find an ostrich, and in Australia, the emu. The plains of La Plata are also home to the agouti and bizcacha – animals with nearly the same habits as hares and rabbits and belonging to the same order of rodents, but with a uniquely American structure. The lofty peaks of the Cordillera are home to an alpine species of bizcacha. The waters do not harbor the beaver or muskrat but the coypu and capybara, rodents of the American type. There are many other examples. The islands off America's coast may differ geologically from the mainland, but their inhabitants are essentially American, even if they are all unique species. In past ages, American types were prevalent on the American continent and in American seas, as discussed in chapter 10. These observations reveal a deep bond between organisms, prevailing through space and time, over the same areas of land and water, and independently of physical conditions. Any naturalist who doesn't inquire what this bond may be lacks curiosity.

According to my theory, the bond is simply inheritance. This is the only cause, as far as we positively know, that produces organisms that are similar to one another. The differences among inhabitants of different regions are attributable to modification through natural selection, and to a much lesser extent the direct influence of physical conditions. The degree of difference depends on the ease with which dominant life forms have migrated, how long ago migrations occurred, the nature and number of immigrants, and their interactions during the struggle for existence. (As previously mentioned, the relationships among organisms are the most important relationships.) Barriers are therefore influential, because they act as checks on migration.

Widely ranging species, abounding in individuals that have already triumphed over many competitors, will have the best chance of seizing new places when they spread into new regions. They will be exposed to different conditions in their new homes and will frequently undergo further modification and improvement; thus they become even more victorious and will generate groups of modified descendants. Inheritance with modification explains why sections of genera, whole genera, and even families are confined to the same areas, as is often the case.

As stated in chapter 10, I do not believe in a law of necessary development. The variability of each species is an independent property, and natural selection can take advantage of it only insofar as it profits the individual in its complex struggle for life. Therefore, the amount of modification in different species is not a uniform quantity. For example, if a number of competing species all migrate to a new and subsequently isolated region, they will be little subject to modification, because migration and isolation cannot, in themselves, effect anything. These principles only become influential by bringing organisms into new relationships with one another and, less significantly, with physical conditions. In chapter 10 I discussed forms that have retained nearly the same characteristics from very remote geological periods; similarly, certain species have migrated over vast spaces without becoming greatly modified.

It is obvious from this discussion that the species within a genus must have proceeded from one source – one ancestor – even though they may inhabit distant parts of the world. It is also obvious that the individual members of a species can be traced back to one spot where their parents were first produced, even if they now inhabit separate and isolated regions. As explained in the previous chapter, it is virtually impossible for identical individuals to stem from distinct species through natural selection.

This brings me to a question that has been discussed extensively by naturalists: have species been created at one point on the earth's surface, or at multiple points? There are, of course, many cases in which it is difficult to understand how the same species could have possibly migrated from one point to many distant isolated points. Nevertheless, the simplicity of the idea that each species was first generated within a single region is captivating. Anyone who rejects it also rejects the true cause of ordinary generation with subsequent migration and calls upon the agency of miracle.

It is universally admitted that in most cases the area inhabited by a species is continuous, and it is considered remarkable or exceptional when a plant or animal inhabits two points so far from each other that the distance cannot be easily traversed by migration. Land mammals are particularly limited in their capacity to migrate across the sea; accordingly, there are no inexplicable cases of the same mammal inhabiting widely separated parts of the world. Geologists understand, for example, that Great Britain was once united with the European mainland and consequently has the same quadrupeds.

If the same species can be generated at two points, why isn't there a single mammal common to Europe and Australia or South America? The environments are nearly the same, so that a multitude of European plants and animals have become naturalized in America and Australia. Moreover, some of the original plants are identical in these distant points of the northern and southern hemispheres. The answer, I believe, is that mammals have not been able to migrate, whereas plants have migrated across vast and broken landscapes through their varied means of dispersal. The significant influence that barriers have had on distribution only makes sense with the deduction that the majority of species produced on one side of a barrier cannot go over to the other side. A few families, many subfamilies, and very many genera and sections of genera are confined to a single region. Also, several naturalists have observed that genera containing very closely related species are usually confined to one area. It would be a strange anomaly if one step lower in the series – at the level of individuals belonging to a species – a different rule prevailed, with species being generated in two or more places instead of being local.

It seems to me – and to many other naturalists – that the most probable explanation for the observed patterns is that each species is generated in only one area and then spreads as far as its powers of migration and subsistence under past and present conditions allow. There are undoubtedly many cases in which we cannot explain how the same species could have passed from one point to another, but geographical and climatic changes, which have certainly occurred within recent geological times, must have interrupted the once continuous ranges of many species. We are reduced to consider whether the exceptions to "continuity of range" are numerous and serious enough to warrant dismissal of the idea altogether, even though general considerations make it probable. It would be hopelessly tedious to discuss all the exceptions, and I do not pretend that explanations could be offered in many cases. However, I will supply some preliminary remarks and then discuss the most striking observations – namely, that the same species inhabit summits of distant mountain ranges, and at distant parts of the Arctic and Antarctic. In chapter 12 I will discuss the broad distribution of freshwater organisms and the fact that the same terrestrial species inhabit mainland areas and islands separated by hundreds of miles of open ocean. If the existence of the same species at distant and isolated points can usually be explained by species having migrated from a single birthplace, then this view – especially when considered with our ignorance of former geological and climatic changes, and of the various occasional means of transport – can safely be accepted as a universal rule.

In discussing this subject I will also consider an equally important question: can the distinct species of a genus, which, according to my theory, have descended from a common ancestor, have migrated from the area inhabited by their ancestor while simultaneously undergoing modification? If the inhabitants of a region are closely related to the species of a second region and it can be demonstrated that there has probably been migration between them, then my theory will be strengthened. The principle of descent with modification explains why the inhabitants of one region should be related to the inhabitants of another region from which it has been stocked. For example, a volcanic island formed a few hundred miles from a continent will probably receive a few colonists from the mainland; their descendants will become modified but will still plainly be related by inheritance to the inhabitants of the continent. Cases like this are common and are inexplicable on the theory of independent creation. This view of the relationship between species in one region to those in another is similar to that recently proposed in an ingenious paper by Mr. Wallace, if the word "variety" is replaced by "species." He concludes that "every species has come into existence coincident both in space and time with a pre-existing closely allied species."

I now know from our correspondence that he attributes this coincidence to generation with modification.

These remarks are not *directly* relevant to a related question: do all the individuals of a species descend from a single pair, a single hermaphrodite, or, as suggested by some authors, from many simultaneously created individuals? According to my theory, a species with members that never mate or cross with others (if such exist) must have descended from a succession of improved varieties that never blended with other varieties but supplanted one another so that at each stage of modification and improvement, all the individuals of each variety will have descended from a single parent. However, in the majority of cases – that is, with all organisms that mate to reproduce or often intercross – the many individuals of a species change simultaneously. This is because intercrossing keeps individuals nearly uniform during the process of modification; thus the total amount of modification at each stage will not derive from a single ancestor. To illustrate what I mean, consider that English racehorses differ slightly from the horses of every other breed, yet their difference and superiority are not a consequence of descent from any single pair, but of continued care in selecting and training many individuals over many generations.

Before discussing the three sets of observations I have selected as showing the greatest problems with "single centers of creation," I must briefly address "means of dispersal." Sir Charles Lyell and others have treated this subject extensively, and I will provide only a short summary. Changes in climate have a powerful influence on migration. A region can be a high road for migration under one climate and impassable under another. Changes in the level of the land are also influential. A narrow isthmus separates two marine faunas; submerge it and the faunas blend. Where the sea now extends, land may have once connected islands or possibly even continents, allowing the movement of land animals. Geologists agree that great changes of level have occurred within the period of existing species. Edward Forbes asserts that all Atlantic islands must have recently been connected with Europe or Africa, and Europe must have recently been connected with America. Other authors have in this way hypothetically bridged every ocean and united almost every island to some mainland. By Forbes's argument, almost all islands were recently connected to a continent, cutting the Gordian knot of dispersal of one species to widely separated points and removing many difficulties. To the best of my judgment, however, invoking such enormous geographical changes within the period of existing species is unjustified. Evidence abounds for large oscillations of the level of continents, but not for huge changes in their positions and extensions so as to have united them within the recent period to one another and to intervening oceanic islands. However, many islands that might have served as halting places for plants and many animals during migration are now submerged beneath the seas. I believe that in coral-producing oceans, sunken islands are now marked by rings of coral, or atolls, standing over them. When one day it is accepted that each species has proceeded from a single birthplace, and when we know something definite about the means of distribution, we will be able to safely speculate about the former extension of continents. But I do not think it will ever be proven that continents that are now separated were, in the recent period, continuous with one another and oceanic islands.[2]

The striking contrast between marine faunas on the opposite shores of most continents, the close relationship between Tertiary inhabitants and modern inhabitants of several regions and seas, a certain degree of relation between the distribution of mammals and the depth of the sea

2. [Alfred Wegener developed the theory of continental drift in the early 1900s to elegantly explain centuries of accumulated geological data, but he could not propose a mechanism to explain why continents should move relative to one another. His ideas were highly controversial and precipitated a revolution in geology not resolved until the 1960s, when plate tectonics came of age. Geophysicists now understand that the earth's crust consists of oceanic and continental plates, all adrift on a semifluid mantle, and jostling one another throughout the eons. Sir Charles Lyell was technically incorrect in some of the processes he proposed to be acting on earth's surface ("oscillations of level," for example), but he did have the significant insight that the physical earth changes drastically, and often very slowly. In this way, Lyell, Darwin, and Wegener belong to the same intellectual tradition. Fortunately for Darwin's ideas, continental drift happens to have produced many of the same effects on the distribution of life forms as Lyell's geology would have. Especially as this part of geology was so poorly understood, Darwin's observations and reflections are remarkably astute. Indeed, in one of his first published writings, he proposed a theory for the formation of coral islands and atolls that is still considered correct. – D.D.]

(as discussed later), as well as other observations concerning distribution do not, it seems to me, support the notion that prodigious geographical revolutions have occurred within the recent period, as required by Forbes's views and espoused by many of his followers. In addition, the nature and relative proportions of the inhabitants of oceanic islands show that they were not once continuous with continents. Their universally volcanic composition suggests that they are not wrecks of sunken continents; if these islands were originally mountain ranges on land, then at least some of them would be composed of granite, metamorphic schists, old fossil-rich rocks, or other such rocks found on mountain summits, instead of consisting of mere piles of volcanic matter.

I also want to briefly discuss what is usually known as "accidental" means of distribution – although it should really be called "occasional" means of distribution. I will discuss only plants. Botanical works list this or that plant as "ill adapted for wide dissemination," but almost nothing is known about how well plants can spread across the sea. Until I tried a few experiments with Mr. Berkeley's aid, it was not even known how long seeds could resist seawater. To my surprise I found that out of eighty-seven kinds of seeds, sixty-four germinated following an immersion of 28 days, and a few survived an immersion of 137 days. For convenience I used mostly small seeds without capsule or fruit, and all of these sank in a few days, which means they could definitely not be floated across wide stretches of ocean even if undamaged by seawater. I then tried some larger fruits, capsules, and so forth, and some of these floated for a long time. Seasoned timber is more buoyant than green timber, and it occurred to me that floods could wash down plants or branches, which could then dry on banks before getting washed into the sea by a fresh rise in the stream. So I dried the stems and branches of ninety-four plants with ripe fruit and placed them in seawater. Most sank quickly, but some that floated for only a short time while green floated much longer when dry. For example, fresh ripe hazelnuts sank immediately, but dried hazelnuts floated for 90 days and germinated after being planted. An asparagus plant with ripe berries floated for 23 days; dried it floated for 85, and the seeds germinated afterward. Ripe seeds of *Helosciadum* sank in two days; dried they floated for more than 90 and germinated afterward. Altogether, out of the ninety-four dried plants, eighteen floated for more than 28 days, and some of these eighteen floated for much longer. So to sum up the whole experiment, sixty-four of eighty-seven seeds germinated after 28 days of immersion, and eighteen of ninety-four dried plants with ripe fruit floated for more than 28 days (note that some of the species in the ripe fruit experiment differed from those in the initial seed immersion experiment). If anything can be inferred from these scanty data, it's that 14 percent of the plants from any region may float along ocean currents for 28 days and retain the ability to germinate. According to Johnston's *Physical Atlas*, the average rate of Atlantic currents is 33 miles per day (with some as fast as 60 miles per day). At this average rate the seeds of 14 percent of the plants from any region could float across 924 miles of sea to another landmass, and once stranded, if blown to a favorable spot by an inland gale, they could germinate.

M. Martens later tried experiments similar to mine, but his were much better because he placed the seeds in a box in the actual sea so that they were alternately wet and exposed to the air like plants that are actually floating. He tried ninety-eight seeds, mostly different from those I used. He chose many large fruits, as well as seeds from plants that live near the ocean, thereby favoring greater average floatation time and a greater resistance to seawater. On the other hand, he did not dry the plants or branches with the fruit, something that would have caused some of them to have floated much longer. The result was that eighteen out of ninety-eight seeds floated for forty-two days and were then able to germinate. But it seems to me that plants exposed to waves would float for less time than those protected from violent movement as in our experiments, so it may be safer to assume that dried seeds from about 10 percent of the plants in a region could float for nine hundred miles and then germinate. It is interesting that large fruits float longer than small fruits. Plants with large seeds and fruit could hardly spread any other way, and Alph. de Candolle has shown that they generally have restricted ranges.

Seeds can sometimes be transported in another way. Drift timber gets washed up on most islands, even those in the midst of the widest seas. Natives of coral islands in the Pacific procure stones for their tools solely from the roots of drifted trees. (These stones are a valuable royal tax.) I find on examination that when irregularly shaped stones are embedded in the roots of trees, small parcels of earth are frequently also enclosed – and so perfectly that not a particle could be washed away during the longest transport. Indeed, out of one small portion of earth completely enclosed by wood in an oak about fifty years old, three dicotyledonous plants germinated (I am certain of the accuracy of this observation). Likewise, I can show that the carcasses of birds floating on the ocean sometimes escape being immediately devoured, and in the crops of floating birds many kinds of seeds retain their vitality. Peas and vetches are killed by even a few days' immersion in seawater, but to my surprise some peas and vetches taken out of the crop of a pigeon that had floated on artificial saltwater for thirty days germinated.

Living birds are surely effective in dispersing seeds. I could give many facts showing how frequently birds of many kinds are blown by gales to vast distances across the ocean. It is safe to assume that under such conditions a bird's speed is often thirty-five miles per hour, and some authors give much higher estimates. I have never observed nutritious seeds to pass through a bird's intestines, but hard seeds from fruit pass through even a turkey's digestive system undamaged. Over the course of two months I found in my garden twelve types of seed from the excrement of small birds; they seemed normal, and when I planted them, some of them germinated. The crops of birds do not secrete gastric juice, and I know by trial that they do not inhibit a seed's ability to germinate. After a bird devours a large amount of food, not all of the grains pass into the gizzard for twelve or even eighteen hours. In this interval a bird can easily be blown five hundred miles; hawks seek tired birds and the contents of their prey's torn crop can thus get scattered. Mr. Brent informs me that one of his friends had to give up flying carrier pigeons from France to England because hawks on the English coast destroyed so many on their arrival. Some hawks and owls eat their prey whole and twelve to twenty hours later disgorge pellets that contain seeds capable of germination, as I know from experiments done in the Zoological Gardens. Some seeds of oat, wheat, millet, canary, hemp, clover, and beet germinated after twelve to twenty-one hours in the stomachs of different birds of prey. Two beet seeds germinated after being retained for two days and fourteen hours. I have noticed that freshwater fish eat seeds of many land and water plants, and fish are often eaten by birds, so the seeds might be transported this way. I forced many kinds of seeds into the stomachs of dead fish and then fed them to fishing eagles, storks, and pelicans. After many hours these birds either rejected the seeds in pellets or passed them in excrement; several of these seeds germinated. Certain seeds, however, were always killed by this process.

Although the beaks and feet of birds are generally quite clean, I can show that soil sometimes does adhere to them. In one case I removed twenty-two grains of dry clay, including a pebble as large as a vetch seed, from one foot of a partridge. So seeds may occasionally be transported large distances this way because soil is almost everywhere charged with seeds. Reflect for a moment on the millions of quails that annually cross the Mediterranean; the earth stuck to their feet must sometimes contain a *few* small seeds. I will come back to this subject.

Icebergs are sometimes loaded with earth and stones and have even carried brushwood, bones, and the nest of a land bird. As suggested by Sir Charles Lyell, they must occasionally have transported seeds within Arctic and Antarctic regions and, during the glacial period, within regions that are now temperate. The Azores have many plants in common with Europe in comparison to other oceanic islands nearer the mainland, and as Mr. H. C. Watson has remarked, these plants are somewhat northern in characteristics for that latitude, so the Azores might have been partly stocked by ice-borne seeds during the glacial period. At my request, Sir Charles Lyell wrote to M. Hartung to inquire whether he had observed any erratic boulders on these islands; Hartung reported finding large fragments of granite and other rocks that do not occur on the archipelago. Thus, it looks as though icebergs unloaded their rocky burdens on the shores of these mid-ocean islands, and it's at least possible that they also brought seeds of northern plants.

Given that these and other yet-to-be-discovered means of dispersal have been in action year after year, for centuries and tens of thousands of years, it would be marvelous if many plants had not become widely spread. These means of transport are sometimes called "accidental," but this is not strictly correct, because ocean currents are not accidental nor is the direction of prevalent winds. Note that very few means of transport can convey seeds very great distances, because seeds do not maintain their viability when exposed to seawater for a great length of time. Nor can they be carried for long in the crops or intestines of birds. However, these modes of transfer are sufficient for occasional transport across several hundred miles of ocean or from a continent to a neighboring island, but not from one distant continent to another. The floras of widely separated continents cannot mingle extensively by these means, so they remain distinct. The currents are such that they cannot bring seeds from North America to Britain, though they may and do bring seeds from the West Indies to Britain's western shores. But seeds thus arriving in Britain cannot endure the climate even if they survive the long immersion in saltwater. Almost every year, one or two land birds are observed to have been blown from North America to the western shores of Ireland and England. These wanderers could only transport seeds by dirt sticking to their feet, which is in itself a rare event. Even if it did happen, the chances of a seed falling on favorable soil and coming to maturity are very small. However, it would be a major error to argue that just because a well-stocked island like Great Britain has not received immigrant species in recent centuries through occasional means of dispersal (as far as we know, this has not happened, and in any case it would be extremely difficult to prove), a poorly stocked island farther from the mainland cannot receive colonists by similar means. I appreciate that out of every twenty seeds or animals transported to an island – even one significantly less stocked than Great Britain – perhaps just one would be well fitted enough to its new home to become naturalized. But this is not a valid argument against the potential effects of occasional means of dispersal over geological time scales – for example, while an island is being upheaved, and before it becomes fully inhabited. On almost bare land, with no destructive insects or birds, nearly every seed that happened to arrive would germinate and survive.

Identical plants and animals on mountain summits separated by hundreds of miles of lowlands, where alpine species could not possibly exist, are a striking case of "the same species at distant points," because migration is apparently not possible across the lowland barriers. It's amazing that so many of the same plants live within the snowy regions of the Alps or Pyrenees and in the very north of Europe, but it's even more amazing that plants on the White Mountains of the United States are all the same as those in Labrador and nearly all the same (as reported by Asa Gray) as those on the loftiest European peaks. Even as long ago as 1747, such observations led Gmelin to conclude that species must have been independently created at multiple locations. This belief would yet endure had not Agassiz and others called attention to the glacial period, which supplies a simple explanation of these facts.

There is evidence of almost every conceivable kind – organic and inorganic – that central Europe and North America were under an arctic climate within a very recent geological period. The ruins of a house burned down by fire do not tell their tale more plainly than the mountains of Scotland and Wales, with their scored flanks, polished surfaces, and perched boulders telling of icy streams that filled their valleys not so long ago. The climate in Europe has changed so much that gigantic moraines left by glaciers in northern Italy are now clothed by vine and maize. Across a large portion of the United States, erratic boulders along with rocks scoured by icebergs and coastal ice reveal a former cold period.

The influence of the glacial climate on Europe's inhabitants has been explained with remarkable clarity by Edward Forbes, whose arguments are substantially as follows. As the cold spread, each successively more southern zone became habitable to arctic organisms and inhabitable to temperate organisms, which became supplanted. The inhabitants of temperate regions simultaneously traveled south unless blocked by barriers, in which case they perished. Snow and ice enveloped mountains, and their alpine inhabitants descended to the plains. By the time the cold peaked, a uniform arctic fauna and flora covered the central parts of Europe as far south as the Alps and Pyrenees and even stretching into Spain. The now-temperate regions of the United States were also covered by arctic plants and animals that were nearly the same as those in Europe. (Current circumpolar organisms are very uniform around the world, and they are assumed to have spread southward in unison.) If the glacial period came on a little earlier or later in North America than in Europe, the migration south also happened slightly earlier or later, but this does not affect the final result.

As the warmth returned, arctic forms retreated north, closely followed by the inhabitants of temperate regions. As the snow melted from the bases of the mountains, the arctic forms ascended higher and higher while their brethren were on a northward journey. When the warmth had fully returned, the same arctic species that had previously lived together as a body on the lowlands of the Old and New Worlds were left isolated on mountain summits, having been exterminated at lower altitudes, and in the arctic regions of both hemispheres.

This explains how identical plants can exist in areas as remote as the mountains of the United States and Europe. It also explains why alpine plants of each mountain range are closely related to arctic forms living due north, or nearly due north; the migration as cold fell, and the retreat as warmth returned, would generally have been along north-south lines. For example, Mr. H. C. Watson remarks that the alpine plants of Scotland are particularly related to the plants of northern Scandinavia, as are the plants of the Pyrenees, as noted by Ramond. Plants of the United States are related to those of Labrador, and plants on the mountains of Siberia are related to those of the arctic regions of Siberia. These ideas are grounded on the well-established occurrence of a glacial period; they illuminate the present distribution of alpine and arctic organisms of Europe and America so well that if the same species are found to inhabit separated mountain peaks in other regions, then even without other evidence I would conclude that a colder climate permitted their migration across low intervening tracts that have since become too warm for their existence.

If at any point since the glacial period the climate has been warmer than it is now,[3] arctic and temperate organisms will have moved a little farther north before returning to their present homes. However, I have not encountered any convincing evidence for this intercalated warm spell sometime after the glacial period.

Arctic forms followed and therefore experienced almost the same climate during their migration south and return north. They also stayed together as a unit, so their relationships were not much disturbed. In accordance with the principles instilled by this book, this means they were not liable to much modification. The situation is different for alpine

3. As propounded by some geologists in the United States based mainly on the distribution of *Gnathodon* fossils.

organisms left isolated by the returning warmth. It is unlikely that all the same arctic species were left on widely separated mountain ranges and have survived there ever since. They also probably mingled with ancient alpine species that must have existed on the mountains before the start of the glacial period. (These older alpine species were temporarily driven down to the plains during the coldest portion of the glacial period.) They will also have been exposed to somewhat different climatic influences. Their relationships will therefore have been a bit disturbed, and they will consequently have become liable to modification. This is in fact the case: in a comparison of current alpine plants and animals of great European mountain ranges, many of the species are identical, but some have formed varieties and a few are distinct but closely related species.

In illustrating what I believe happened during the glacial period, I have assumed that when the period started, arctic forms were as uniform around polar regions as they are today. However, the remarks I made earlier about distribution apply not only to strictly arctic forms but also to many subarctic and some northern temperate forms; some of these are the same on the lower mountains and on the plains of North America and Europe. So it's reasonable to ask: how do I account for the necessary amount of similarity among subarctic and northern temperate forms around the world at the start of the glacial period? Subarctic and northern temperate species of Europe and America are currently separated from one another by the Atlantic Ocean and by the northern part of the Pacific Ocean. During the glacial period, they must have been still more effectively separated by wider spans of ocean while inhabiting more southerly regions than they do today. This problem can be resolved by invoking still earlier climatic changes of an opposite nature. There is evidence that during the newer Pliocene period (which was before the glacial period), the climate was warmer than it is now, yet the majority of the world's species were the same as they are today. Thus, organisms now living at 60° of latitude probably lived at latitude 66°–67° (just under the polar circle) during the Pliocene. Arctic organisms lived on broken land still closer to the pole. Look at a globe and notice that there is almost continuous land under the polar circle, from Western Europe through Siberia to eastern America. I attribute the necessary amount of uniformity in subarctic and northern temperate organisms after the glacial period to this continuity of circumpolar land and the freedom it lends to inter-migration under a favorable climate.

Because the continents have long remained in nearly the same relative positions, though subjected to large partial oscillations in level, I would extend the above view and infer that during some earlier and still warmer period, such as the older Pliocene, many of the same plants and animals inhabited the almost continuous circumpolar land. In both the Old and New Worlds, these plants and animals slowly migrated south as the climate became colder, long before the glacial period. Their descendants, mostly in a modified condition, now inhabit central parts of Europe and the United States. This explains the relationship between North American and European organisms; very few organisms are identical, and the relationship is remarkable considering the distance between the two areas and their separation by the Atlantic Ocean. This also sheds light on the singular fact mentioned by several observers that European and American organisms from the later stages of the Tertiary period were more closely related to one another than they are today. During those warmer periods, the northern parts of the Old and New Worlds were almost continuously united by land, serving as a bridge for intermigration. This bridge has since been rendered impassable by cold.

Species migrating south of the polar circle during the increasing cold of the Pliocene period were immediately cut off from one another. As far as more temperate organisms are concerned, this separation occurred long ago. As the plants and animals moved south, one set mingled with native American organisms and had to compete with them, and the other encountered the organisms of the Old World. All conditions favorable to modification were met; conditions favorable to far more modification than with Alpine organisms left isolated during a much more recent period on mountain ranges and arctic regions. This is why there are very few identical species now living in temperate regions,[4] but in every great class there are many forms ranked variously by naturalists as "geographical races" or "distinct species" and a host of closely related or representative forms ranked by all naturalists as "distinct species."

4. Asa Gray has recently shown that more plants are identical than was previously thought.

As on land, so in the waters of the sea there was a slow southern migration of marine fauna that had once been nearly uniform along the shores of the polar circle during the Pliocene or even earlier. When considered with the theory of modification, this accounts for the many closely related forms now living in areas completely sundered, the existing and Tertiary representative forms on the eastern and western shores of temperate North America, and the still more amazing case of many closely related crustaceans, some fish, and other marine animals in the Mediterranean and the seas of Japan (as described in Dana's impressive work) – areas now separated by a continent and by nearly a hemisphere of equatorial ocean.

These groups of related but not identical organisms separated by oceans, and the associations among past and present inhabitants of temperate North America and Europe, are incomprehensible within the theory of creation. They could not have been created alike in correspondence with nearly similar physical conditions; certain parts of South America closely resemble southern continents of the eastern hemisphere in physical condition, yet their inhabitants are completely different.

Returning to the more immediate subject – the glacial period – I am convinced that Forbes's view can be extended. In Europe there is clear evidence of the cold period from the western shores of Britain to the Ural Mountains and south to the Pyrenees. Siberia was similarly affected as inferred from frozen mammals and the nature of mountain vegetation. Glaciers have left marks of their low descent along the Himalaya at points nine hundred miles apart, and Dr. Hooker saw maize growing on gigantic ancient moraines in Sikkim.[5] South of the equator there is direct evidence for a past period of glacial action in New Zealand. The same story is told by New Zealand's plants: the same plants can be found on widely separated mountains. If one account that has been published can be trusted, then there is also direct evidence for glacial action in the southeastern corner of Australia.

Turning to America, ice-borne rock fragments have been observed on the eastern side of the northern half of the continent as far south as latitudes 36°–37°, and on the shores of the Pacific, where the climate is

5. [A region of northern India in the eastern Himalaya. – D.D.]

now so different, to latitude 46°. Erratic boulders have been noticed in the Rocky Mountains. Glaciers once extended far below their present latitude in the Cordillera of equatorial South America. In Central Chile I was astonished at the structure of a vast eight-hundred-foot-high mound of detritus crossing a valley of the Andes. I am now convinced that it is a gigantic moraine left far below any existing glacier. Farther south on both sides of the continent, from latitude 41° to the southernmost tip, huge boulders transported far from their source give clear evidence of the former action of glaciers.

Whether the glacial period was exactly simultaneous at these various points on opposite sides of the world is unknown, but there is good evidence in almost every case that it happened within the most recent geological period. There is also excellent evidence that it lasted for an enormous time, as measured in years, at each point. The cold may have come on or ceased earlier at one point on the globe than another, but because it endured for so long at each point and occurred simultaneously in a geological sense, it was probably simultaneous throughout the world for at least some time. Without distinct evidence to the contrary, it can at least be claimed that glacial action was probably simultaneous on the eastern and western sides of North America, in the Cordillera under the equator and under the warmer temperate zones, and on both sides of the southern extremity of South America. With this admitted, it is difficult to avoid concluding that the temperature of the whole world was simultaneously cooler during this period. However, it would suffice for my argument if the temperature was lower along certain broad belts of longitude at the same time.

On this view of the whole world, or at least of broad longitudinal belts, having been simultaneously colder from pole to pole, much about the present distribution of identical and allied species makes sense. Dr. Hooker has shown that between forty and fifty of Tierra del Fuego's flowering plants – no small portion of this region's scanty flora – are common with Europe, despite the huge distance between the two regions. There are also many closely related species. The lofty mountains of equatorial America host many peculiar species belonging to European genera. Gardner found a few European genera on the highest mountains of Brazil absent in the wide intervening hot regions. Similarly, Humboldt long ago found species belonging to genera characteristic of the Cordillera on the Silla of Caraccas. Several European forms and a few representatives of the peculiar flora of the Cape of Good Hope occur on the mountains of Abyssinia.[6] The Cape of Good Hope is home to a very few European species, believed not to have been introduced by humans, and a few representative European forms that have not been discovered in the tropical parts of Africa are found on the mountains. Many plants occur on the Himalaya, isolated mountain ranges of India, heights of Ceylon, and volcanic cones of Java that are either identical or represent one another – and at the same time represent plants of Europe – yet are not found in the intervening hot lowlands. A list of the genera collected on the loftier peaks of Java conjures a picture of a collection made on a hill in Europe! Even more striking is that southern Australian forms are clearly represented by plants growing on the summits of Borneo. I hear from Dr. Hooker that some of these Australian forms extend along the heights of the Malayan Peninsula and are also thinly scattered over India and as far as Japan. Dr. F. Müller has discovered several European species on the southern mountains of Australia; other species not introduced by humans occur on the lowlands; and I am informed by Dr. Hooker that many European genera can be found in Australia but not the intervening hot and dry zones. Dr. Hooker supplies analogous facts about the plants of New Zealand in his *Introduction to the Flora of New Zealand.* Throughout the world, plants growing on lofty mountains and on temperate lowlands of the northern and southern hemispheres are sometimes identical, but more often they are distinct species that are remarkably related.

This summary applies only to plants, but there are analogous observations concerning the distribution of terrestrial animals. There are similar cases with marine organisms as well. For example, I can quote a remark made by that highest authority, Professor Dana: "It is certainly a wonderful fact that New Zealand should have a closer resemblance in its crustacea to Great Britain, its antipode, than to any other part of the world." Sir J. Richardson also speaks of northern fish types on the shores of New Zealand, Tasmania, and other places. Dr. Hooker informs me

6. [Modern-day Ethiopia. – D.D.]

that twenty-five species of algae are common to New Zealand and to Europe but have not been found in tropical seas in between.

Note that northern species and forms found in the southern parts of the southern hemisphere and on mountain ranges of the tropical zone are not arctic but northern temperate. Mr. H. C. Watson recently remarked, "In receding from polar towards equatorial latitudes, the Alpine or mountain floras really become less and less arctic." Many forms living on mountains in warm regions and in the southern hemisphere are of doubtful value, because some naturalists rank them as distinct species and some as varieties. But some of them are definitely identical, and many must be ranked as unique species even though they are closely related to northern forms.

What light can be thrown on these observations by the view – supported by a large body of geological evidence – that the whole world, or a large part of it, was simultaneously much colder during the glacial period than it is today? The glacial period lasted a long time (as measured in years). Given that vast spaces have been invaded by some naturalized plants and animals in the past few mere centuries, the glacial period would have afforded ample time for any amount of migration. As the cold slowly came on, all the tropical plants and other organisms retreated from both sides toward the equator, followed by temperate organisms, and these by arctic organisms. The tropical plants probably suffered a great deal of extinction. No one can say how much; maybe the tropics supported as many species as are now crowded together on the Cape of Good Hope and in parts of temperate Australia. Many tropical plants and animals can withstand a considerable amount of cold; therefore, many might have escaped extermination during a moderate fall in temperature, especially by escaping to the warmest spots. The important thing to keep in mind is that all tropical organisms suffered to some extent. After migrating nearer to the equator, temperate organisms will have suffered less, despite the somewhat new conditions. (Many temperate plants can withstand a much warmer climate than their own if protected from the inroads of competitors.) Given that the tropical organisms were in a suffering state and could not have presented a firm front against intruders, it is possible that some of the more vigorous temperate forms penetrated the native ranks and reached or even crossed the equator. Of course the invasion would have been greatly favored by high land and perhaps by a dry climate. (Dr. Falconer informs me that it is the damp along with the heat of the tropics that is so destructive to perennial plants from a temperate climate.) The most humid and hottest regions, however, will have offered an asylum for the tropical natives. The mountain ranges northwest of the Himalaya and the long line of the Cordillera seem to have afforded two great lines of invasion. Strikingly, all the flowering plants common to Tierra del Fuego and Europe (about forty-six) still exist in North America,[7] which must have fallen on the line of march. Some temperate species entered and crossed even the lowlands of the tropics when the cold was most intense – when arctic forms had migrated about twenty-five degrees of latitude from their native region and covered the land at the foot of the Pyrenees. During this extreme cold the climate at the equator at sea level was about the same as the climate at the equator now at an elevation of six or seven thousand feet. Large spaces of the tropical lowlands were probably clothed by mingled tropical and temperate vegetation like that now growing with strange luxuriance at the base of the Himalaya, graphically described by Dr. Hooker.

So this, I believe, is how a considerable number of plants, a few terrestrial animals, and some marine organisms migrated during the glacial period from the northern and southern temperate zones to the tropics; some even crossed the equator. As the warmth returned, temperate forms naturally ascended mountains as they were exterminated on lowlands. Those that had not reached the equator migrated back north or south to their former homes. Forms that had crossed the equator (mostly northern forms) traveled still farther from their homes to the temperate latitudes of the opposite hemisphere. Although paleontological evidence suggests that arctic shelled mollusks barely underwent any modification during their long journey south and return north, the situation may have been entirely different for those intruding forms that settled on tropical mountains and in the southern hemisphere. They were surrounded by strangers and had to compete with many new life forms. Selected modifications in their structures and habits would have been profitable. Although many of these wanderers are still clearly related to their northern

7. Recently communicated to me by Dr. Hooker.

and southern brethren by inheritance, they are now varieties or distinct species in their new homes.

Apparently, many more identical plants and related forms migrated from the north to the south than in the opposite direction.[8] (However, there are a few southern vegetable forms on the mountains of Borneo and Abyssinia.) This asymmetry probably resulted from the greater extent of land in the north and the greater number of northern forms and their consequent advancement through natural selection and competition to a higher stage of perfection, or dominating power, than southern forms. When the two groups mingled during the glacial period, the northern forms were enabled to beat the less powerful southern forms. The same phenomenon is observable today in the many European organisms covering the ground of La Plata, and to a lesser degree of Australia, which to a certain extent have beaten those that are native. In contrast, very few southern organisms have become naturalized in any part of Europe, though hides, wool, and other objects likely to carry seeds have been imported to Europe during the last two or three centuries from La Plata, and during the last thirty or forty years from Australia. Something similar must have happened on tropical mountains, as before the glacial period they were stocked with endemic alpine forms, but almost everywhere these have yielded to more dominant forms generated in the larger areas and more efficient workshops of the north. On many islands, invasive organisms equal or even outnumber endemic species, and if the natives have not already been exterminated, then their numbers have been greatly reduced – the first stage of extinction. A mountain is an island on land, and tropical mountains before the glacial period must have been completely isolated. I believe that the productions of these islands on land yielded to those produced in the larger areas of the north just as the productions of actual islands have recently yielded to invasive species introduced by humans.

I am far from supposing that this view resolves all difficulties about the ranges and relationships of related species living in northern and southern temperate zones and on tropical mountains. Many difficul-

8. Dr. Hooker strongly insists on this observation with respect to America, and Alph. de Candolle does so with respect to Australia.

ties remain to be solved. I do not pretend to know the exact lines and means of migration, the reasons why certain species migrated and others didn't, or why certain species have been modified and given rise to new groups while others remained unaltered. These observations cannot be explained until we first understand why certain species and not others become naturalized in foreign lands, or why one species ranges twice or thrice as far as another.

Some of the most interesting unsolved difficulties are clearly stated by Dr. Hooker in his botanical works on Antarctic regions. I cannot discuss them here, but I will say that the occurrence of identical species in regions so enormously remote as Kerguelen Land, New Zealand, and Fuegia can be explained if they were dispersed by icebergs at the close of the glacial period, as suggested by Lyell. However, the existence of distinct species from exclusively southern genera at these and other distant points of the southern hemisphere is a much greater difficulty for my theory of descent with modification. Some of these species are so distinct that there has not been enough time since the start of the glacial period for their migration and subsequent modification to the necessary degree. These observations suggest that peculiar and very distinct species have migrated in radiating lines from some common center. As with the northern hemisphere, I am inclined to propose a former warm period in the southern hemisphere before the start of the glacial period when Antarctic lands now covered with ice supported a unique isolated flora. Before this flora was destroyed by the glacial period, a few forms were widely dispersed to various parts of the southern hemisphere by occasional means of dispersal, with the aid of now sunken islands as stepping-stones. Forms may have spread at the start of the glacial period on icebergs. These mechanisms have tinted the southern shores of America, Australia, and New Zealand with the same peculiar forms of plant life.

In a striking passage, Sir Charles Lyell has speculated, with language almost identical to mine, on the effects of great climatic changes on geographical distribution. The world has recently felt one of his great cycles of change. This view, combined with the principle of modification through natural selection, explains many observations of the current distribution of identical and related life forms. The living waters flowed from the north and the south for one short period, crossing at the equator, and flowing with greater force from the north so as to inundate the south. The tide of life thus left its living drift on mountain summits, in a line gently rising from the arctic lowlands to a great height at the equator. The various organisms left stranded this way are like the indigenous human groups driven up and surviving in the mountains of almost every region, serving as a record, full of interest to us, of the former inhabitants of surrounding lowlands.

12

THE GEOGRAPHICAL DISTRIBUTION
OF LIFE, CONTINUED

LAKES AND RIVER SYSTEMS ARE SEPARATED FROM ONE another by barriers of land, so it might seem obvious that freshwater species do not range across many bodies of water and that because the sea is an even greater barrier, they would never spread to distant regions. But actually the reverse is true. Not only do many freshwater species from quite different classes have enormous ranges, but interrelated species prevail throughout the world! When I first collected freshwater insects, shelled mollusks, and other freshwater species in Brazil, I remember being surprised at their similarity to those in Britain, and at the dissimilarity of the surrounding terrestrial organisms.

The capacity of freshwater organisms to range widely may be unexpected, but in most cases it can be explained by their being adapted to make short and frequent migrations from pond to pond or stream to stream – a tendency for wide dispersal follows as an almost necessary consequence. I can here consider only a few cases. With respect to fish, I believe that the same species never occurs in freshwater on widely separated continents. But within one continent, individual species range widely and almost capriciously: two river systems can have some fish in common and some different. A few observations favor the possibility that they occasionally spread by "accidental" means; for example, live fish are sometimes dropped by whirlwinds in India, and fish eggs remain viable after their removal from water. However, I attribute the dispersal of freshwater fish mainly to minor changes in land level within the recent period that caused rivers to flow into one another. Examples could also be given of this result brought about by floods without any changes in level. The loess of the Rhine reveals considerable changes in level within a very recent geological period when the surface was inhabited by still-extant terrestrial and freshwater shelled mollusks. The remarkable difference between fish on opposite sides of continuous mountain ranges, which must have parted river systems and prevented them from intertwining since an early period, suggests the same conclusion.

The many cases of related freshwater fish occurring in widely separated parts of the world are currently inexplicable. But some freshwater fish belong to very ancient forms, and in these cases there has been ample time for great geographical changes, and as a consequence, for much migration. Also, saltwater fish can slowly become accustomed to living in freshwater, and according to Valenciennes, there are very few fish groups confined exclusively to freshwater. A marine member of a freshwater group could conceivably travel far along the shores of the sea and subsequently become modified and adapted to the freshwaters of a distant land.

Some freshwater shelled mollusks have a very wide range, and species related to them – which, according to my theory, descended from a common ancestor and must have proceeded from a single location – prevail throughout the world. Their distribution perplexed me at first because their eggs are unlikely to be transported by birds, and they perish in seawater, as do the adults. I could not even understand how some naturalized species have spread rapidly *within* a region. But I have made two observations that shed light on this problem (no doubt many others remain to be made). I have twice seen a duck suddenly emerging from a duckweed-covered pond with the little plants adhering to its back. And in moving a little duckweed from one aquarium to another, I unintentionally stocked the second with freshwater shelled mollusks from the first. But another agency is perhaps more effective: I suspended a duck's feet – which may represent those of a bird sleeping in a pond – in an aquarium. The aquarium contained many hatching eggs of freshwater shelled mollusks, and I found hatched mollusks crawling on the duck's feet, clinging so firmly that when they were taken out of the water, they could not be jarred off. (They voluntarily dropped off at a more advanced age.) These just-hatched mollusks survived on the duck's feet, in damp air, from twelve to twenty hours. In this amount of time a duck or heron could fly at least six or seven hundred miles and would surely alight on a pond or rivulet if blown across the sea to an oceanic island or any other distant point. Sir Charles Lyell informs me that a *Dyticus* beetle has been caught with a freshwater mollusk similar to a limpet firmly adhering to it, and a water beetle of the same family once landed aboard the *Beagle* when she was forty-five miles from the nearest land. How much farther it might have flown with a favorable gale no one can tell.

With respect to plants, many freshwater and even marsh species have enormous ranges over continents, even reaching the most remote oceanic islands. As remarked by Alph. de Candolle, this is strikingly illustrated by large groups of terrestrial plants that have only a few aquatic members; these few members always seem to have broad ranges, as if in consequence. Favorable means of dispersal are the explanation. I mentioned previously that soil sometimes adheres to the feet and beaks of birds. Wading birds, which frequent the muddy edges of ponds, are the likeliest to have muddy feet, and birds of this order are the greatest wanderers, occasionally found on remote and barren islands in the open ocean. They are unlikely to alight on the surface of the sea, so the soil would not be washed off; when making land they would fly to their natural freshwater haunts. I don't think botanists are aware of the extent to which pond mud is charged with seeds. I have tried several little experiments, but I will describe only the most amazing one. In February I took three tablespoonfuls of mud from three different points under the water at the edge of a little pond. When dry, this mud weighed only 6.75 ounces. I kept it covered in my study for six months, pulling up and noting each plant as it grew. There were 537 plants of many kinds, and yet the viscid mud was all contained in a breakfast cup! Given this, I think it would be inexplicable if birds *did not* transport the seeds of freshwater plants vast distances, thereby giving the plants large ranges. The same process may spread the eggs of some small freshwater animals.

Other processes, some unknown, probably contribute. As previously mentioned, freshwater fish eat some kinds of seeds, although they reject many other kinds after swallowing them; even small fish swallow seeds of moderate size, like those of the yellow water lily and pondweed. Herons and other birds have been devouring fish daily, century after century. They take flight and go to other waters or are blown across the sea, and seeds retain their power of germination for many hours after being rejected in pellets or in excrement. When I saw the large size of the seeds of the *Nelumbium* water lily and remembered Alph. de Candolle's remarks on the plant, I thought its distribution inexplicable. However, Audubon found seeds of the great southern water lily in a heron's stomach.[1] By analogy, a heron flying to another pond and getting a hearty meal of fish would probably reject from its stomach a pellet containing undigested *Nelumbium* seeds, or the seeds might be dropped by the bird while feeding its young, the same way that fish are sometimes dropped.

In considering these various means of distribution, note that when a pond or stream is first formed – for example, on a rising islet – it is uninhabited, and a single seed or egg has a good chance of succeeding. There is always a struggle for existence between individuals of species already occupying a pond, even if there are only a few. However, in general the number of kinds of aquatic species is small compared with the number of kinds on land, so competition is probably less severe among aquatic than among terrestrial species. Therefore an intruder from the waters of a foreign region has a better chance of seizing a niche than is the case with terrestrial colonists. Some, perhaps many, freshwater organisms are low on the scale of nature, so they probably become modified less quickly than organisms higher on the scale. This gives a longer-than-average amount of time for the migration of a given aquatic species. It is also probable that many species once ranged as continuously as is possible for freshwater organisms to range, thus getting distributed across immense areas before becoming extinct in intermediate regions. But I believe that the wide distribution of freshwater plants and lower animals results mainly from the dispersal of their seeds and eggs by animals, especially freshwater birds, which naturally travel between often distant bodies of water. Like a careful gardener, Nature takes her seeds from one particular bed and drops them into another and equally well fitted one.

The inhabitants of oceanic islands present the last of the three major challenges to the view that all individuals, both within a species and of related species, have descended from a common ancestor and have thus proceeded from a common location to inhabit distant points on the globe. I have already stated that I disagree with Forbes's view on conti-

1. The lily was probably the *Nelumbium luteum,* according to Dr. Hooker.

nental extensions, which, if true, would mean that all currently existing islands were recently joined nearly or completely to continents. This interpretation would remove many difficulties, but it would not address all observations about insular organisms. In the following sections I will not confine myself to just problems of dispersal but will also consider other observations that are relevant to the two theories of independent creation and of descent with modification.

The number of species inhabiting oceanic islands is few compared with the number of species on equal continental areas. (Alph. de Candolle agrees to this for plants and Wollaston for insects.) Compare the mere 750 flowering plants of large New Zealand, extending over 780 miles of latitude and varied in habitat, with those on an equal area of the Cape of Good Hope or Australia; something other than differences in the physical conditions has caused a great difference in number. Even the uniform county of Cambridgeshire has 847 plants, and the little island of Anglesea has 764. (A few ferns and introduced plants are included in these tallies, and in some other respects the comparison is not quite fair.) There is evidence that the barren island of Ascension originally possessed fewer than half a dozen flowering plants, yet many plants have become naturalized there – as they have on New Zealand and on every other oceanic island that can be named. On St. Helena naturalized plants and animals have nearly or completely exterminated many native organisms. Anyone who maintains the doctrine of the creation of each species must also admit that a significant number of the best-adapted plants and animals have not been created on oceanic islands. Humans have unintentionally stocked islands from various other sources more fully and perfectly than nature.

On oceanic islands the number of species is scanty, but the proportion of endemic species – those found nowhere else in the world – is often huge. To appreciate this, compare the number of endemic land snails on Madeira, or the number of endemic birds on the Galápagos Archipelago, with the number found on any continent, taking into account the area of the island with respect to that continent. My theory anticipates this result because, as already explained, species arriving at long intervals to a new and isolated region have to compete with new associates and will be especially liable to modification, thereby often generating groups of modified descendants. However, it does not follow that just because nearly all the species of one class on an island are peculiar, those of another class or subsection of the same class are also peculiar. This distinction seems to depend on the immigration of *groups of species* that do not become modified, so their mutual relationships are maintained. On the Galápagos, nearly every land bird but only two out of eleven marine birds are unique; marine birds can obviously arrive at these islands more easily than land birds. In contrast, Bermuda, which lies at about the same distance from North America as the Galápagos do from South America, and which has a very peculiar soil, does not possess a single endemic land bird. According to Mr. J. M. Jones's account, many North American birds either periodically or occasionally visit Bermuda during their great annual migrations. There are no unique birds on Madeira, and Mr. E. V. Harcourt informs me that many European and African birds are blown there almost every year. Bermuda and Madeira have thus been stocked by birds that for ages had struggled together in their former homes and had become adapted to one another; in their new homes each is kept in its proper place and habitat by the others, so they are not as liable to modification. Madeira is inhabited by an amazing number of land snails, whereas not one marine shelled mollusk is confined to its shores alone. It is unknown how marine shelled mollusks are dispersed, but it is possible that their eggs or larvae can travel across three or four hundred miles of ocean more easily than those of land snails, perhaps attached to seaweed, floating timber, or the feet of wading birds. The insect orders found on Madeira apparently present analogous examples.

Oceanic islands are sometimes deficient in certain classes of organisms, so their niches are apparently occupied by other inhabitants. Mammalian niches are occupied by reptiles on the Galápagos and by giant wingless birds on New Zealand. With respect to plants, Dr. Hooker has shown that proportional numbers of the different orders are very different on the Galápagos from what they are elsewhere. Such cases are usually accounted for by the physical conditions on the islands, but this is unlikely. Facility of immigration has been at least as important.

Many interesting little facts could be given about the inhabitants of remote islands. For example, certain islands devoid of mammals have some endemic plants with beautifully hooked seeds, even though few adaptations are more striking than hooked seeds for transport by the wool and fur of mammals. This case is not a difficulty for my theory, because a hooked seed can be transported to an island by other means. The plant would then become modified while retaining its hooked seeds, transforming into an endemic species with as useless an appendage as any rudimentary organ (like the shriveled wings under the soldered armor of many insular beetles). Islands often have trees or bushes belonging to orders that elsewhere contain only herbaceous species. As Alph. de Candolle has shown, trees generally have confined ranges, for whatever reason, so they are unlikely to reach oceanic islands. An herbaceous plant has no chance of successfully competing in stature with a fully developed tree, but established on an island and in competition with only other herbaceous plants, it may be advantageous to become taller and taller, rising above the other plants. If so, natural selection would tend to increase the stature of herbaceous plants growing on an island regardless of the order they belong to, converting them first into bushes and then into trees.

Bory St. Vincent long ago remarked that batrachians (the frogs, toads, and newts) have never been found on any of the islands studding the great oceans. I have investigated this assertion and found it to be true. (I have been assured that a frog exists on the mountains of New Zealand, but I suspect that this exception, if true, can be explained by glacial agency.) The general absence of frogs, toads, and newts on so many oceanic islands cannot be accounted for by physical conditions. In fact, islands seem particularly suitable for these animals: frogs have been introduced to Madeira, the Azores, and Mauritius only to have multiplied so much as to have become a nuisance. My theory explains why, despite being suited to island life, batrachians are not usually found there: the animals and their spawn are immediately killed by seawater, so they are unlikely to spread across the sea to islands. But according to the theory of creation, it would be very difficult to explain why they should not have been created there.

Mammals provide another, similar case. I have not found a single definite example of a terrestrial mammal, excluding domesticated animals kept by natives, inhabiting an island more than three hundred miles from a continent or continental island.[2] Many islands situated at lesser

---

2. I have carefully examined records from even the oldest voyages, although I have not yet completed my research.

distances are equally barren. The Falkland Islands, which are inhabited by a wolflike fox, come nearest to an exception. However, these islands are not oceanic, because they lie on a bank connected with the mainland. Also, icebergs once delivered boulders to its western shores and might have also transported the foxes, as frequently happens now in arctic regions. Small islands are certainly *capable* of supporting small mammals, because in many parts of the world these animals can be found on small islands near continents. There are almost no islands on which small mammals haven't become naturalized and greatly multiplied. It cannot be claimed, on the common notion of creation, that there has not been enough time for the creation of mammals on islands, because many volcanic islands are indeed ancient, as shown by the stupendous erosion they have undergone and by their Tertiary strata. There has also been enough time for the generation of endemic species belonging to other classes. And on continents, mammals are thought to appear and disappear faster than other, lower animals.

Though terrestrial mammals do not occur on oceanic islands, aerial mammals do. New Zealand has two bat species found nowhere else in the world, and Norfolk Island, the Viti Archipelago, the Bonin Islands, the Caroline and Marianne Archipelagos, and Mauritius all harbor unique bats. Why would the supposed creative force produce bats but no other mammals on oceanic islands? My theory easily addresses this matter: no terrestrial mammal can be transported across a wide space of ocean, but bats can fly across. Bats have been seen wandering by day far over the Atlantic Ocean, and two North American species either regularly or occasionally visit Bermuda, which is six hundred miles from the mainland. Mr. Tomes has specially studied this family, and he informs me that many species have enormous ranges and are found both on continents and on distant islands. Supposing only that such wandering species have been modified through natural selection in their new homes relative to their new environments explains the presence of endemic bats on islands in the absence of all terrestrial mammals.

Distance between islands and continents is not the only factor that determines whether terrestrial mammals will be present on an island. Whether we find the same or related species of mammal on an island as compared with the nearest mainland also depends on the depth of the sea that separates the two, and this is largely independent of distance. Mr. Windsor Earl has made some striking observations about this with respect to the Malay Archipelago, which is traversed near Celebes by a space of deep ocean separating two very distinct mammalian faunas. On either side the islands are situated on moderately deep submarine banks and are inhabited by identical or related quadrupeds. A few anomalies probably occur on this large archipelago, and it is difficult to judge some cases, because certain mammals have probably been introduced by humans, but the admirable zeal and research of Mr. Wallace will soon illuminate the natural history of this archipelago. I have not yet followed up on this subject with respect to all other parts of the world, but as far as I have gone, the relation generally holds. Britain is separated from Europe by a shallow channel, and the mammals are the same on both sides. Many islands separated from Australia by similar channels are analogous cases. But the West Indian Islands stand on a deeply submerged bank, nearly one thousand fathoms in depth, and although these islands are home to American forms, the species and even the genera are distinct. The amount of modification always depends to some extent on the amount of elapsed time, and during changes in level it is obvious that islands separated by shallow channels are more likely to have recently been connected to the mainland than islands separated by deep channels. This explains the relationship between the depth of the sea between island and mainland and the strength of the affinity between their respective mammalian inhabitants – a relationship that is inexplicable on the view of independent acts of creation.

All of these observations concerning the inhabitants of oceanic islands agree more closely with the view that occasional means of dispersal have been efficient over long periods of time than with the view that all oceanic islands have recently been connected to the nearest continent. If the second case were correct, migration would probably have been more complete and organisms would be more equally modified, in accordance with the crucial importance of the relationships among organisms.

I do not deny that there are many serious difficulties to understanding how some of the inhabitants of remote islands have reached their present homes, regardless of whether they still maintain their original form or have been modified since their arrival. However, the probability of now nonexistent islands having served as stepping-stones must not be overlooked, and I will provide one example. Almost all oceanic islands, even the most isolated and diminutive of them, are inhabited by land snails; these are generally endemic species.[3] Land snails are easily killed by salt, and their eggs (at least the ones I have tested) sink in seawater and are killed by it. But according to my theory, there must be some unknown but highly efficient means for their transport. Could the just-hatched young occasionally crawl on and adhere to the feet of birds roosting on the ground? This would be one means of transport. And here's another: when a land snail hibernates, a membrane forms over the mouth of its shell. In this state it may be floated in chinks of drifting timber over moderately broad areas of the sea. I found that while they are hibernating, several species can withstand immersion in seawater for seven days. I tried an additional experiment with *Helix pomatia;* when it hibernated again, I immersed it in seawater for twenty days, and it recovered completely. As this species has a thick calcareous operculum,[4] I removed it, and when it formed a new membranous covering, I immersed it again for fourteen days. It recovered and crawled away, but more experiments are needed to study this phenomenon.

The most striking and important observation concerning the inhabitants of oceanic islands is their relationship to species on the mainland without actually being the same species. There are many examples, and I will use the Galápagos Archipelago, which lies at the equator and between five hundred and six hundred miles from the shores of South America. Here, almost every inhabitant bears the unmistakable stamp of the American continent. There are twenty-six land birds, and twenty-five of them are ranked by Mr. Gould as distinct species, supposed to have been created on the islands. But the close relationship between most of these birds and American species in every characteristic, habit, gesture, and song is manifest. The same applies to other animals, and to nearly all the plants as shown by Dr. Hooker in his memoir on the flora of the Galápagos. The naturalist looking at the inhabitants of these volcanic

3. Dr. Augustus A. Gould gives several interesting cases of land snails on Pacific islands.
4. [The operculum is the "door" used by a snail to close its shell. – D.D.]

islands feels as though he were standing on American land. Why should species supposed to have been created on the Galápagos and nowhere else bear so clear a stamp of relationship to those created in America? There is nothing about the environment, geological nature, height, climate, or proportions of associated classes that closely resembles the South American coast. In fact there are considerable dissimilarities. Conversely, there is a considerable resemblance between the volcanic soils, climates, heights, and sizes of the Galápagos Islands and Cape de Verde Islands, yet there is an entire and absolute difference between their inhabitants! The inhabitants of the Cape de Verde Islands are related to those of Africa, just as the inhabitants of the Galápagos are related to those of America. This grand fact can find no explanation whatsoever through the common notion of independent creation. With my theory, though, it is obvious that the Galápagos should receive colonists from America, whether by occasional means of dispersal or across formerly continuous land, and the Cape de Verde Islands should receive colonists from Africa. These colonists would be liable to modifications, inheritance betraying their original birthplace.

Many analogous observations could be given. In fact, it's an almost universal rule that the endemic organisms of an island are related to the inhabitants of the nearest continent or other nearby islands. The exceptions are few and most can be explained. The Kerguelen Islands are closer to Africa than America, and yet as Dr. Hooker has shown, the islands' plants are very closely related to those of America. This anomaly disappears if the Kerguelen Islands have mainly been stocked by seeds brought with earth and stones on icebergs that drifted on the prevailing currents. New Zealand's endemic plants are more closely related to those of Australia, the nearest mainland, than any other region, as expected. But the flora is *also* obviously related to that of South America; this is the next nearest continent, but it is so enormously far that the observation becomes an anomaly. Yet this too almost disappears as a difficulty if New Zealand, South America, and other southern lands were partially stocked long ago from a mutually distant point – namely, the Antarctic islands, when they were clothed with vegetation before the glacial period. Dr. Hooker assures me that although the affinity between the flora of the southwestern corner of Australia and the Cape of Good Hope is weak, it is real; this is a far more remarkable case and currently inexplicable. However, the relationship is confined to plants, and I am sure it will someday be explained.

The close relationship between specifically distinct inhabitants of an island and the nearest continent is sometimes displayed on a small scale within the limits of a single archipelago. The islands of the Galápagos are, marvelously, inhabited by very closely related species, as I have discussed elsewhere. The inhabitants of each island are mostly distinct but are incomparably more closely related to one another than to the inhabitants of any other place in the world. My theory anticipates precisely this because the islands are so close to one another that they should almost certainly have received colonists from the same original source, or from each other. But the dissimilarities between the endemic inhabitants of the islands can be used to argue against my theory: if these islands are within sight of one another and have the same geological nature, height, climate, and so forth, then why have the colonists been modified differently (though the differences are slight)? This problem occurred to me long ago, and it stems mainly from the deep-seated misconception that the physical conditions of a region are the most important element for its inhabitants when in fact the nature of the other inhabitants with which each has to compete is generally far more crucial to success. On the Galápagos there is a considerable amount of difference from island to island between inhabitants also found in other parts of the world. (Endemic species, for the moment, cannot be included, because I want to consider how they have come to be modified since arrival.) This difference is expected if the islands came to be inhabited through occasional means of dispersal. For example, the seed of one plant could have been brought to one island and the seed of another plant to another island. When an immigrant form settles on one or more islands, or when it subsequently spreads from one island to another, it is exposed to different environments because it has to compete with different sets of organisms. For example, a plant would encounter the best soil more perfectly occupied by distinct species on one island than another. It would also be exposed to attacks from somewhat different enemies. If the plant then varied, natural selection would favor different varieties on different islands. However, some species might spread but retain the same characteristics throughout the group, just as on continents some species spread extensively but remain the same.

The most surprising thing about this case of the Galápagos, and to a lesser degree about some other, similar cases, is that the new species formed on the separate islands have not quickly spread across the archipelago. Although the islands are within sight of one another, they are separated by deep arms of the sea, in most cases wider than the British Channel. There is no reason to assume that the islands were once continuous. Sea currents are swift and sweep across the archipelago, and gales are rare. The islands are, therefore, far more effectively separated from one another than they appear to be on a map. Nevertheless, they have many species in common, both those found elsewhere in the world and those confined to the archipelago. It can be inferred from certain observations that they probably spread from one island to the others. But I think we often misunderstand the probability that closely related species will invade each others' territories when put into free intercommunication. Of course, if one species has any advantage over the other, it will wholly or partially replace it in a brief time. But if both are equally well fitted to their own niches, they will probably hold their own places and keep separate for almost any length of time. Many species naturalized by humans have spread astonishingly quickly over new regions, but this does not mean that most species would spread this way. Organisms that become naturalized in foreign regions are not closely related to the native inhabitants; they are very distinct species often belonging to different genera, as shown by Alph. de Candolle. Even many of the birds on the Galápagos are distinct on each island, despite being adapted to flying between islands. For example, there are three closely related mockingbird species each confined to its own island. Suppose that the mockingbird of Chatham Island is blown to Charles Island, which has its own mockingbird species. Can the newcomer establish itself? Charles Island is densely inhabited by its own species because more eggs are annually laid there than can possibly be reared, and the mockingbird unique to Charles Island is at least as well fitted for *its* home as the species unique to Chatham Island. Sir Charles Lyell and Mr. Wollaston have communicated to me some interesting information that is relevant to this subject. Madeira and the adjoining islet of Porto Santo possess many distinct but representative land snails, some of which live in stone crevices. Although large quantities of stone are transported from Porto Santo to Madeira, Madeira has not become colonized by the Porto Santo species. Both islands have been colonized by European land snails, which presumably have advantages over the native species. Given these considerations, it's actually not surprising that the endemic and representative species of the Galápagos have not spread to all the islands. In many other cases, like those involving multiple regions on the same continent, previous occupation has probably played an important part in preventing the commingling of species in equivalent conditions. For example, the southeastern and southwestern corners of Australia have nearly the same physical environments and are united by continuous land, yet they are home to a vast number of distinct mammals, birds, and plants.

Inhabitants of oceanic islands are identical or clearly related to the inhabitants of the region from which colonists could most readily have arrived; after immigrating, colonists are modified and better fitted to their new homes. This principle applies throughout nature – on every mountain, in every lake and marsh. Alpine species are related to those of the surrounding lowlands (with the exception of those organisms that spread throughout the world during the recent glacial period, mostly plants). Alpine hummingbirds, rodents, plants, and other species of South America are all strictly American forms. This makes sense because as a mountain gets slowly upheaved, it will naturally be colonized from the surrounding lowlands. The same applies to inhabitants of lakes and marshes, taking into account that in these cases easy dispersal has given the same general forms to the whole world and to the blind cave inhabitants of America and Europe. Additional, analogous observations could be given. I predict that whenever two regions – regardless of the distance between them – harbor many closely related species, there will always be a few identical ones. This follows from there having been migration between the two regions, as described above. And wherever many closely related forms occur, some naturalists will rank them as distinct species while others will rank them as varieties; these doubtful forms demonstrate the steps in the process of modification.

The relationship between the power and extent of migration of a species (either in the present or in the past under different physical conditions) and the presence of related species in widely separated locations is also shown in another and more general way. Mr. Gould remarked to me long ago that in those genera of birds that range across the world, many of the species also have very wide ranges. This rule is probably generally true, but it would be difficult to prove. Among mammals it can clearly be observed with bats, and less so with the cat family and dog family and by comparing the distribution of butterflies and beetles. It applies to most freshwater organisms, many genera of which range over the world with many individual species having enormous physical ranges. I do not mean that in a world-ranging genus all the species have a wide range or even that they range widely on average, but that some of the species will range widely. The facility with which widely ranging species vary and give rise to new forms will largely determine their average range. For example, if two varieties of the same species inhabit America and Europe, it means they have a huge range. However, if variation had been a little greater, the two varieties would be considered two species, and the common range would be much smaller. Neither does the rule state that a species capable of crossing barriers and ranging widely – like certain powerfully winged birds – actually will. A wide range implies not only an ability to cross barriers but also the more important power of being victorious in the struggle for life in distant lands with foreign associates. The view that all the species of a genus, though now distributed across the world, have descended from a single ancestor predicts that at least some of the species will range widely, as is found to be the case. It is *necessary* for the unmodified parent to range widely, undergoing modification while diffusing, and to find itself under diverse conditions favorable to the conversion of its offspring into new varieties and, ultimately, new species.

In considering the wide distribution of certain genera, recall that some are very ancient and must have branched off from a common ancestor long ago. In such cases there will have been ample time for great climatic and geographical changes, "accidental" dispersal of species, and the migration of some of the species to all regions of the world, where they may have become slightly modified to their new conditions. There is geological evidence that species low on the scale of a great class usually change more slowly than higher forms, so lower forms are more likely to range widely while retaining the same specific characteristics. Many lower forms also produce tiny eggs and seeds that are well suited to extensive transport. These tendencies probably account for the observation recently discussed by Alph. de Candolle with respect to plants: the lower a group of organisms is, the more likely it is to range widely.

All the relationships just discussed are completely inexplicable through the common notion of the independent creation of each species but *are* explicable through colonization from the nearest and most likely source coupled to subsequent modification and adaptation of the colonists to their new homes.

In this and the last two chapters I have attempted to demonstrate that due allowance must be made for our ignorance about the full effects of all the changes in climate and land level that must have happened within the recent period, and also of other similar transformations which may have occurred. Very little is known of the many unusual "occasional means of dispersal," a subject that has barely been studied experimentally. Also, species may often have ranged continuously over an area and then become extinct in some tracts. These considerations mitigate challenges to the idea that all the individuals of a given species, regardless of their location, have descended from the same parents. This conclusion, designated by many naturalists as "single centers of creation," stems from general considerations, especially from the importance of barriers and the distribution of subgenera, genera, and families.

Difficulties with the view that species belonging to the same genus must have spread from one parental source can also be addressed if the same allowances for our ignorance are made as before. Additionally, recall that some life forms change slowly, allowing great lengths of time for their migration.

To show how climatic change can affect distribution, I discussed the profound influence of the modern glacial period. (The glacial period affected the whole world, or at least great meridional belts, simultaneously.) To show the diversity of "occasional means of dispersal," I discussed the many ways freshwater organisms can be dispersed.

If it is right that in the long course of time the individuals of the same species and of allied species have proceeded from a single source, then all the major elements of geographical distribution can be explained through migrations, generally of dominant organisms, coupled with subsequent modification and multiplication of new forms. This explains the importance of land and water barriers separating zoological and botanical provinces. It explains the localization of subgenera, genera, and families. It explains the similarities among inhabitants of plains and mountains, forests, marshes, and deserts at different latitudes of a given continent, as well as the similarity of the current inhabitants to the extinct species that once inhabited the same continent.

Two areas with nearly the same physical environments are often inhabited by very different life forms. This makes sense because the relationships among organisms are of the highest importance. And thus the differences accord with how much time has passed since new organisms entered a region; with the way in which certain forms but not others entered, and in what numbers; with whether newcomers came in direct competition with original inhabitants; and with whether the newcomers had the possibility of varying quickly. Consequently, different regions have infinitely diverse environments independently of physical conditions, because organisms act and react to one another almost endlessly. This predicts highly modified groups, slightly modified groups, and large and small groups. And this is what is found in the major geographical regions of the world.

These principles explain why oceanic islands have few but mostly endemic species, and why differences in migratory ability mean that one group can have all of its species endemic while another group even within the same class can have its species common to other parts of the world. They explain why oceanic islands lack representatives of certain entire groups – like frogs and terrestrial mammals – while even the most isolated islands possess their own unique bats. They explain the relationship between the presence of mammals and the depth of the sea between an island and the mainland. They explain why all the inhabitants of an archipelago are related to one another but why each islet harbors its own distinct species, with the inhabitants also related, only not as closely, to those of the nearest continent or to some other source that probably supplied migrants. They explain the correlation between the presence of identical species, varieties, and distinct but representative species in two areas even if they are far apart.

The late Edward Forbes often insisted on a parallelism to the laws of life throughout time and space, the rules governing the succession of forms in the past being nearly the same as those governing the differences among organisms in different areas at the present. The endurance of each species and group of species is continuous in time.[5] It is a general rule that the area inhabited by a single species or group of species is continuous. There are many exceptions, but I have attempted to show that they can be explained through migration during some earlier period when conditions were different, or through occasional means of dispersal and the species becoming extinct in intermediate tracts. Species and groups of species have points of maximum development both in time and space. Groups of species belonging either to a certain time or certain area are often characterized by minor similarities, as in shape or color. Looking across a long succession of time or to faraway regions of the world in the present, some organisms differ slightly and others from a different class, order, or even only a different family differ greatly. The lower members of each class generally change less than the higher throughout time and space, but there are exceptions in both cases. My theory renders these relationships across time and space intelligible. Forms of life that have changed across time in one part of the world and those that have changed after migrating great distances are connected by the same bond of ordinary reproduction. The more closely any two forms are related in blood, the more closely they will stand in time and space. In both cases the laws of variation are the same, and modifications have been accumulated by the same power of natural selection.

5. The exceptions are so few that they can be attributed to paleontologists not yet having discovered forms that are absent in some intermediate deposit but present in those above and below.

13

AFFINITIES BETWEEN ORGANISMS: MORPHOLOGY, EMBRYOLOGY, AND RUDIMENTARY ORGANS

FROM THE FIRST DAWN OF LIFE, ALL ORGANISMS HAVE resembled one another in descending degrees, so they can be classed into groups subordinate to groups. This classification is not arbitrary like the grouping of stars into constellations. Its meaning would have been straightforward if one group had been fitted exclusively to land and another to water, one to feed on flesh, another on vegetation, and so on. The reality of nature is very different, with members of even the same subgroup commonly having different habits. In chapters 2 and 4, on variation and natural selection, respectively, I attempted to show that widely ranging, highly diffused, and common species, dominant within large genera, vary the most. The varieties – *the insipient species* – produced in this way ultimately become new and distinct species, and these tend to produce other new and dominant species through inheritance. As a consequence, large groups, generally with many dominant species, tend to continue increasing indefinitely. I also tried to demonstrate that the varying descendants of each species strive to occupy as many differ-ent places as possible in the environment, resulting in a constant ten-dency for their characteristics to diverge. This conclusion is supported by studying the great diversity of life forms competing closely in any small area and by considering certain observations about naturalization.

I also attempted to show that organisms that are increasing in num-ber and diverging in characteristics tend to supplant and exterminate the less divergent and less improved preceding organisms. Return to the diagram illustrating these principles in chapter 4, and see how the modi-fied descendants proceeding from a common ancestor become broken up into groups subordinate to groups. Each letter on the uppermost line could be taken to represent a genus composed of several species; all the genera on this line collectively form a class because they have descended from one ancient but undepicted ancestor and have therefore inherited something in common. By this logic, the three genera on the left have a lot in common and constitute a subfamily distinct from the subfamily immediately to the right, which is made up of two genera diverged from a common ancestor at the fifth stage of descent. Taken together, these five genera also share some commonalities and form a family distinct from the one composed of the three genera still further to the right, which diverged even earlier. All the genera descended from (A) form an order, and the genera descended from (I) make another, distinct order. So the diagram represents many species descended from a single ancestor grouped into genera that are part of, or subordinate to, subfamilies, families, and orders, all united into one class. This fully explains natural history's grand feature of groups ordered under groups, which is so familiar that it does not always strike us.

Naturalists try to arrange the species, genera, and families of each class in a so-called natural system. But what does this system mean? Some authors consider it merely a scheme to arrange together living things that are alike and separate those that are different. Others consider it an artificial system to describe general characteristics as briefly as possible; for example, a single sentence will give all the characteristics common to mammals, another will give those common to carnivores, another those common to the dog genus, and by adding one more sentence a full description can be given of each kind of dog. This is undeniably an ingenious and useful system, but many naturalists think it means something more: that it reveals the plan of the Creator. But unless they specify what this plan means – order in time and space, for example – it adds nothing to our knowledge. There is a famous expression by Linnaeus, often encountered in a somewhat concealed form, that characteristics do not make the genus, but the genus defines the characteristics. Such comments suggest that classification includes something beyond mere resemblance. I believe something else *is* included: classification partially reveals kinship of descent, the only known cause of similarity among organisms, hidden by various degrees of modification.

Reflecting on the rules followed in classification reveals the difficulties of assuming that classification gives some unknown plan of creation or that it is simply a scheme for describing characteristics and placing together forms that are most alike. In ancient times it was thought that parts of structure that determine an organism's habits and its niche are very important for classification, but this is entirely false. No one considers the external similarity of a mouse to a shrew, of a dugong to a whale, or of a whale to a fish, as important. Although these characteristics are intimately connected with the life of the organism, they are ranked only as "analogical" characteristics. (I will return to the topic of these resemblances.) It could even be made a general rule that the less any part of organization is relevant to special habits, the more important it is for classification. For example, in discussing the dugong, Owen writes, "The generative organs being those which are most remotely related to the habits and food of an animal, I have always regarded as affording very clear indications of its true affinities. We are least likely in the modifications of these organs to mistake a merely analogical for an essential character." Similarly with plants, it is remarkable that the vegetative organs on which their whole life depends are of little significance to their classification, except in the first main divisions, but their reproductive organs and their seeds are of paramount importance!

So resemblances among parts of the organization important to a being's welfare cannot be trusted for classification. This may be the reason why almost all naturalists stress resemblances among organs of vital or physiological importance, an interpretation that usually applies. The relevance of these organs to classification depends on their constancy throughout large groups of species, and this in turn depends on their having been subjected to relatively little change as the species were adapted to their environments. It can almost be demonstrated conclusively that mere physiological value does not determine classificatory value by observing that an organ with nearly the same physiological value in related groups has very different classificatory value. No naturalist can study any group without noticing this, and the writings of almost every author acknowledge it fully. It will suffice to quote the highest authority – Robert Brown – who in writing of certain organs in the Proteaceae states that their generic importance, "like that of all their parts, not only in this but, as I apprehend, in every natural family, is very unequal, and in some cases seems to be entirely lost." In another work he writes that the Connaraceae genera "differ in having one or more ovaria, in the existence or absence of albumen, in the imbricate or valvular aestivation. Any one of these characters singly is frequently of more than generic importance, though even when all taken together they appear insufficient to separate *Cnestis* from *Connarus*." An example from among insects: in one large division of the Hymenoptera, the antennae are constant in structure – as Westwood has remarked. In another division they differ greatly and the differences are not very valuable in classification, yet the antennae in these two divisions of the same order are equally important physiologically. There are many other examples.

Similarly, rudimentary or degenerate organs serve no crucial physiological role, but they are often valuable in classification. Rudimentary teeth in the upper jaws of young ruminants and certain rudimentary leg bones are very useful in demonstrating the relationship between ruminants and pachyderms. Robert Brown asserts that rudimentary florets are paramount to the classification of grasses.

Many characteristics are universally accepted as useful in defining whole groups but are derived from parts of trivial physiological importance. Examples include whether or not there is an open passage between the nostrils and mouth (according to Owen, this is the only characteristic that definitively distinguishes fish and reptiles); the inflection of the angle of the jaws in marsupials; the way an insect's wings fold; color in certain algae; hair on parts of the flower in grasses; dermal covering (e.g., hair, feathers) in vertebrates. If the duck-billed platypus were covered with feathers instead of hair, this minor external characteristic would be as important an aid to naturalists in determining this strange creature's relationship to birds and reptiles as the structure of any vital internal organ.

The importance of trivial characteristics for classification depends mainly on their correlation to other characteristics. (The value of an aggregate of characteristics is evident in natural history.) As often remarked, a species may be different from related species in several prevalent characteristics of physiological importance and yet leave no doubt where it should be classified. This is why classification founded on just one characteristic always fails regardless of how important that characteristic may be: no part of organization is universally constant. The value of an aggregate of characteristics, even when none is individually important, explains Linnaeus's saying that characteristics do not define the genus, but the genus defines the characteristics. This maxim seems founded on an appreciation of many trivial resemblances, each too slight for definition. Certain plants of the Malpighiaceae bear both perfect and degraded flowers. A. de Jussieu has remarked that in the degraded flowers, "the greater number of the characters proper to the species, to the genus, to the family, to the class, disappear, and thus laugh at our classification." Over the course of several years, the *Aspicarpa* genus in France produced only degraded flowers, which departed wonderfully in several of the most important structural characteristics from the proper type of the order; yet M. Richard sagaciously saw, as Jussieu observes, that *Aspicarpa* should remain in the Malpighiaceae. This case illustrates the spirit with which classifications are sometimes necessarily founded.

On a practical level, when naturalists are at work they do not bother about the physiological value of the characteristics used in defining a group or in allocating a particular species. If a characteristic is common to many organisms and nearly uniform, it is assigned a high value; if a characteristic is common to fewer organisms, it takes on a subordinate value. This principle has been declared broadly true by some naturalists, notably the excellent botanist Augustin St. Hilaire. If certain characteristics are always correlated even though no apparent bond can be discovered between them, they are assigned special value. Important organs like those for propelling or aerating the blood, or for propagating the species, are nearly uniform in most animal groups and are very useful for classification, but in some groups all of these vital organs offer characteristics of subordinate value.

It is understandable that embryonic characteristics are as important as those of the adult, because classification embraces all ages of each species. But it is not obvious, based on common notions, why the structure of the embryo should be more important for classification than the structure of the adult, which alone interacts fully with the environment. Yet the great naturalists Milne Edwards and Agassiz assert that embryonic characteristics are the most important for animal classification, and this doctrine has generally been accepted as correct. It also applies to flowering plants, the two main divisions of which are founded on embryonic characteristics: the number and position of embryonic leaves and the mode of plumule and radicle development. The discussion of embryology will reveal why such characteristics are so valuable given that classification tacitly includes the idea of descent.

Classifications are often obviously influenced by chains of affinities. It is easy to list characteristics common to all birds, but such a list has been impossible to define for crustaceans. There are crustaceans at opposite ends of the series with almost no characteristic in common, but because these are obviously related to others, and these to others, and so on, they can be recognized as crustaceans and not members of some other articulate class. Geographic distribution is often used in classification, though perhaps not quite logically. It has been especially applied to very large groups of closely related forms. Temminick asserts the utility or even necessity of this practice for certain groups of birds, and it has been followed by several entomologists and botanists.

Finally, the comparative value of the various groupings of species (orders, suborders, families, subfamilies, genera) seems almost arbitrary, at least at present; several of the best botanists, such as Mr. Bentham, insist on this. There are examples among plants and insects of a group first ranked by naturalists as a genus and then raised to the subfamily or family category, not because further research revealed important structural differences, but because several related species with slightly different graded differences were discovered.

Unless I have greatly deceived myself, all of these rules, aids, and difficulties in classification are explained by the view that the natural system is founded on descent with modification. Characteristics seen by naturalists to signify a true link between two or more species have been inherited from a common ancestor, and in this light all true classification is genealogical. Naturalists have been unconsciously seeking the community of descent, not some unknown plan of creation, or a system to describe general characteristics, or the mere grouping of the similar and separation of the dissimilar.

I should explain what I mean more fully. To be natural, the arrangement of the groups within each class, properly ordered relative to other groups, must be genealogical, but the amount of difference in the groups or branches may differ significantly due to varying amounts of modification even though they are equally related to a common ancestor. This variation in the amount of difference is expressed by forms being ranked in different genera, families, sections, or orders.

Referring back to the diagram in chapter 4, let (A) through (L) represent related genera during the Silurian period descended from a common ancestor from an unknown earlier period. Modified descendants of species from genera (A), (F), and (I) exist today, represented by the fifteen genera on the top line (a^{14} through z^{14}). All of these modified descendants stem from a single species, so they can be called "cousins" to the millionth degree. They are represented as equally related in descent, yet they differ greatly from one another. Forms descended from (A) are now broken up into two or three families and constitute a distinct order from the descendants of (I), also broken up into two families. Genus (A)'s existing descendants cannot be ranked in the same genus as the ancestor (A), and the same applies to (I) and (I)'s descendants. F^{14}'s descent involves only slight modification, so it is ranked together with the ancestral genus F, just as some extant species belong to Silurian genera.

So the amount or value of the differences between organisms that are related to the same degree is highly variable.[1] Nevertheless, their genealogical arrangement is accurate in the present *and* at each successive period of descent. All the modified descendants of (A) or (I) inherited something in common from their common ancestor. However, if any of (A)'s or (I)'s descendants have become modified so extensively as to completely lose traces of their ancestry, their place in natural classification is lost, as seems to have occurred with some existing organisms. The line of descent from genus (F) proceeded with little modification but defines a single genus occupying its correct intermediate position, despite its isolation. Genus (F) was originally intermediate between (A) and (I), and the genera descended from these two have to some extent inherited the ancestral characteristics.[2] The natural system is genealogi-

---

1. ["Organisms that are related to the same degree" means they are descended from a common ancestor. – D.D.]

2. I have attempted to show this natural arrangement on paper, in the diagram, but it's too simplistic. It would have been even more difficult to depict a natural arrangement without a branching diagram. It is exceptionally difficult to represent the relationships among the organisms of a group on a flat surface in a series, with the group names written out linearly.

cally arranged, like a pedigree, because of descent with modification. But to convey the varying amounts of modification undergone by different groups, they are ranked under so-called genera, subfamilies, families, sections, orders, and classes.

This interpretation of classification is illustrated by the relationships among languages. A genealogical arrangement of the human races from a hypothetical perfect pedigree of mankind would provide the best classification of languages spoken throughout the world. This would be the *only* possible arrangement if all extinct languages and all intermediate and slowly changing dialects were included. Some very ancient languages may have changed little, giving rise to few new languages, while others changed extensively, giving rise to many new languages and dialects through the spread and subsequent isolation and variable civilization of human races – all descended from a common race. Degrees of difference in languages from a common source would have to be expressed as groups subordinate to groups, but the proper arrangement, or even the only possible one, would still be genealogical; it would describe the natural connections between all extinct and modern languages, and the filiation and origin of each tongue.

As a confirmation of this view, consider the classification of varieties thought or known to have descended from one species. Subvarieties are grouped under varieties, and varieties are grouped under species. Several other grades are required for domesticated organisms, as discussed with pigeons. The origin of "groups subordinate to groups" is the same for varieties *and* species: closeness of descent with various degrees of modification. The rules of classification are nearly the same for both. Authors have insisted on classifying varieties according to a natural system instead of an artificial one. For example, they caution that two varieties of pineapple should not be grouped together just because their fruit happens to be nearly identical, even though the fruit is the most important part. Swedish and common turnips are not grouped together despite their similar thick and edible stems. Whatever part is most constant is used in classifying varieties; the great agriculturalist Marshall says that horns are useful in classifying cattle because they are less variable than body shape or color, or other characteristics. With sheep, horns are less useful because they are less constant. If a real pedigree of varieties were available, then a genealogical classification – as attempted by some authors – would be universally preferred. Inheritance keeps related forms together, regardless of the amount of modification. Some tumbler pigeon subvarieties differ crucially from others in having a long beak, yet they are kept together because of their common tumbling habit. Every naturalist has actually invoked descent in classifying organisms in the wild, because the two sexes are grouped together as one species, even though they sometimes differ enormously in important characteristics. The adults of certain barnacle males and hermaphrodites are not alike, yet they are never categorized separately. Naturalists treat the multiple larval stages of an individual as one species regardless of how much they differ from one another and from the adult, just as they do for the alternate generations characterized by Steenstrup, and these can be considered the same individual only in a technical sense. Naturalists include organisms with monstrosities as part of a species; they also include varieties, not only because they resemble the ancestral form but also because they descended from it. Naturalists who believe that the cowslip descended from the primrose, or vice versa, rank them as a single species and give only a single definition. As soon as the three orchid forms *Monochanthus*, *Myanthus*, and *Catasetum* – once ranked as distinct genera – were found to sometimes be produced on the same spike, they were immediately ranked as a single species. In light of this system, what should naturalists do if it could be proven that one species of kangaroo descended by a long course of modification from a bear? Should this one species be ranked with bears? What should be done with the other species? Of course it's preposterous: what would be done if a kangaroo gave birth to a bear? According to all analogy, it would be ranked with bears, but then all the other kangaroo species would be grouped under the bear genus – again, preposterous. Wherever there has been close common descent, there will also be close resemblances or relationships.[3]

---

3. [Darwin's argument from common descent implies that through inheritance very different organisms will be directly related through time, linked by many generations, and this lineage of individuals could in principle be defined as a single entity. In fact, by this logic, all life on earth throughout history could be defined as a single species, but that wouldn't be a very useful definition. – D.D.]

Common descent is universally used to group together individuals of the same species, even though the males and females and larvae are sometimes very different. It is also used to organize varieties, which have undergone modification – sometimes significantly so. The same element of descent has been used unconsciously in grouping species under genera and genera under higher groups, although in these cases the modification has been greater and taken longer to complete. This is the only way to make sense out of the rules and guidelines followed by the best systematists. There are no pedigrees, so community of descent must be discerned using resemblances of any kind. Those characteristics are chosen that are least likely to have been modified relative to the environment recently faced by each species; rudimentary structures are therefore as good as, or sometimes even better than, other parts. A characteristic may be minor – mere inflection of the angle of the jaw, the way an insect's wings fold, whether the skin is covered with hair or feathers – but if it prevails throughout many different species, and especially if they have different habits, it assumes high value, because its prevalence can only be explained through inheritance from a common ancestor. A conclusion based on a single point of structure may be incorrect, but when several even minor characteristics occur together throughout a large group of organisms with different habits, then by the theory of descent it is certain that they have been inherited from a common ancestor. It is these correlated or aggregated characteristics that have special value in classification.

A species or group of species is often safely classed with related forms even if it differs in the most important characteristics, so long as a sufficient number of other qualities – even unimportant ones – betray the hidden bond of descent. And if two forms do not have a single characteristic in common but a chain of intermediate groups implies their community of descent, they are put in the same class. Special value is attached to organs with high physiological importance – those that serve to preserve life under diverse conditions – because they are generally the most constant, but if such an organ is found to differ significantly in another group or section of a group, it is valued less in classification. (I will later demonstrate why embryological characteristics are so important for classification.) Geographic distribution is sometimes useful in classing large and widely distributed genera, because all species of a given genus inhabiting any distinct and isolated region have probably descended from a common ancestor.

These concepts illuminate the very important distinction between true affinities and analogical resemblances.[4,5] The resemblance of the body shape and fin-like anterior limbs between the dugong (a pachyderm) and the whale, and between both these mammals and fishes, is analogical. There are countless examples among insects, which is why Linnaeus, misled by external appearances, actually classed a homopterous insect as a moth. Even domestic varieties exhibit a similar phenomenon, as illustrated by the thickened stems of the common and Swedish turnips. The resemblance between greyhounds and racehorses is no more fanciful than the analogies drawn by some authors between very distinct animals. My view that characteristics are important for classification only insofar as they reveal descent explains why analogical characteristics, although very important to an organism's welfare, are almost useless to the systematist. Animals derived from distinct lines of descent can easily become adapted to similar conditions and thereby assume close external resemblances, but these do not reveal – in fact, they tend to conceal – their true lines of descent. It also explains the apparent paradox that characteristics found to be analogical when one class or order is compared with another can give true affinities when members *within* a group are considered: body shape and fin-like limbs are analogical when whales are compared to fishes, being adaptations for swimming in both classes, but they indicate true affinity among members of the whale family, which agree in so many other major and minor characteristics that general body shape and limb structure *must* have been inherited from a common ancestor. The same applies among fishes.

---

4. Lamarck was the first to point out this distinction; Macleay and others have ably followed.

5. [A "true (or real) affinity" is termed a "homology" in modern evolutionary biology. It is a similarity between two species due to common descent. An analogy is a similarity between two species due to convergent evolution: similar structures evolved by two unrelated species to deal with a similar environment. – D.D.]

Species belonging to large genera and descended from dominant species tend to inherit the advantages that made their groups large and their ancestors dominant, so they are likely to spread widely and seize many niches. Large dominant groups therefore tend to increase in size and consequently supplant many smaller and feebler groups. This accounts for all living *and* extinct organisms falling under a few orders, still fewer classes, and one great natural system. In a striking demonstration of how widely spread across the world and how few higher groups are, the discovery of Australia has not added a single insect belonging to a new order, and as I learn from Dr. Hooker, it has added only two or three small orders to the plant kingdom.

In the chapter on geological succession I tried to show that ancient life forms often exhibit characteristics that are intermediate between existing groups because each group has generally diverged significantly during the long process of modification. Occasionally, descendants of intermediate ancestral forms are only slightly modified and constitute the so-called aberrant groups. According to my theory, the more aberrant a form, the greater the number of exterminated connecting forms. There is some evidence for aberrant groups having suffered severe extinctions: they are usually represented by very few species, and even these tend to be distinct from one another (again, implying extinction). The genera of the platypus and lungfish would not be any less aberrant were each represented by a dozen species rather than one, but investigation shows that such richness in species does not typify aberrant genera. This is explained by considering aberrant groups as failing and conquered by more successful competitors, with a few members preserved by some unusual coincidence of favorable circumstances.

Mr. Waterhouse remarks that an animal from one group with affinity to another distinct group often shows a general rather than singular affinity. For example, the bizcacha is the rodent most closely related to marsupials, but the similarities are to the entire marsupial order and not one specific member. In this case the similarities are considered real and not analogical, so they are the result of common descent. Either all rodents – including the bizcacha – branched off from some very ancient marsupial that had characteristics intermediate to existing marsupials, or both groups branched off from a common ancestor and have undergone significant modification in divergent directions. Either way the bizcacha has retained by inheritance more characteristics of its ancestor than other rodents, so it is indirectly related to all or nearly all marsupials, not specially to just one. Mr. Waterhouse also remarks that among marsupials the phascolomys most closely resembles the rodent order. But in this case the resemblance is probably analogical, the phascolomys having become adapted to rodent-like habits. The elder de Candolle makes similar observations about the general affinities among plant orders.

The principle that species descended from some common ancestor multiply and gradually diverge in characteristics while retaining others in common by inheritance illuminates the exceedingly complex and radiating relationships connecting constituent members of a family or higher group. Some modified characteristics of a common ancestor appear in all members of a family – now broken up by extinction into groups and subgroups. The member species are consequently related by complex affinities of various strengths, mounting up through many predecessors (as illustrated in the tree diagram). Naturalists face extraordinary difficulty in organizing the similarities among living and extinct members of a given family, because it is difficult to demonstrate the direct relationships that link them even by the aid of a genealogical tree.

As discussed in chapter 4, extinction plays an important part in defining and widening distinctions between groups in each class. It even accounts for the distinctness of whole classes from one another: birds from other vertebrates, for example, if many ancient life forms that once connected the early ancestors of birds and other vertebrate classes have gone extinct. The extinction of life forms that once connected fish and frogs has been less total; it has been even less in some other classes, like the crustaceans, where the most wonderfully diverse forms are still linked by a long (though broken) chain of similarities.

Extinction separates groups. It does not make them. If every form that has ever existed on earth suddenly reappeared, it would be impossible to define groups, because they would all blend together by gradations as fine as those that separate existing, closely related varieties. Nevertheless, a natural classification, or at least a natural *arrangement*, would be possible. Turning to the diagram, let (A) through (L) represent eleven genera of the Silurian period, some of which produced large groups of modified descendants. Imagine that every intermediate between these genera and their primordial ancestor, and every intermediate in each branch and sub-branch were still alive. It would be impossible to distinguish group members from their more immediate ancestors, or these ancestors from *their* ancient and unknown ancestor, yet the natural arrangement of the diagram would still hold. Inheritance ensures that all of (A)'s descendants retain something in common. (Two branches on a tree may be discrete, but at the fork they blend together.) Although groups could not be defined, types or forms representing most characteristics of each group, small or large, could be picked to give a general idea of the value of the differences between them. We would be driven to this if all the forms that have ever lived throughout time and space in any class were actually collected! We will never succeed in gathering such a perfect collection, although we are moving in that direction with some classes. In a recent paper Milne Edwards asserts the importance of considering "types" whether or not the groups they belong to can be separated.

Natural selection – resulting from the struggle for existence, almost inevitably inducing extinction and divergence among descendants of a dominant ancestral species – explains the salient and universal feature of affinities among organisms: their organization into groups subordinate to groups. The concept of common descent is used in classifying individuals of both sexes and all ages as one species even though they have few characteristics in common. Common descent is used in classifying accepted varieties regardless of how different they are from the parent. *Common descent is the hidden matrix naturalists have sought in the "natural system."* The idea that this system, insofar as it has been perfected, is actually genealogical, expressing grades of difference between descendants of a common ancestor as genera, families, orders, and so on, illuminates the rules naturalists are compelled to use in classification. It explains why some resemblances are more valued than others. It explains why organs that are rudimentary and useless, or of minor physiological importance, can be used in classification. It explains why analogical characteristics are used in comparisons within a group but not between distinct groups. It explains why all living and extinct forms can be linked together in one great system. And it explains how members of each class are linked together by a complex and radiating web of relationships. This web will probably never be disentangled, but with a distinct object in view, rather than some unknown "plan of creation," there is hope of sure, slow progress.

Members of a class, regardless of their habits, resemble one another in morphological organization. "Unity of type," or "homology of parts and organs," is the subject of morphology, the most interesting topic – indeed the soul – of natural history. What is more curious than the homologous pattern of construction of the human hand, formed for grasping; the mole hand, formed for digging; the horse leg; the porpoise paddle; and the bat wing, all containing the same bones in the same relative positions? St. Hilaire asserts the importance of relative connections within homologous organs: constituent parts may change significantly in form and size, but always they remain connected in the same arrangement. For example, the bones of the arm and forearm or thigh and leg are never transposed, which is why the same names can be given to homologous bones in very different animals. The same rule is observed in the construction of insect mouths. The sphinx moth's immensely long, spiral proboscis; the curious folded one on bees or bugs; and the beetle's great jaws are terrifically different, yet all of these organs, serving unique purposes, are formed by infinite modifications to an upper lip, mandibles, and two pairs of maxillae. Analogous rules govern the construction of mouths and limbs in crustaceans and of flowers in plants.

It is hopeless to try explaining this similarity of patterns among members of the same class using the concept of "utility" or "final causes," as Owen expressly admits in his interesting work *Nature of the Limbs: A Discourse.* The common notion that each organism was independently created allows us to say only, "So it is. It has so pleased the Creator to construct each plant and animal." But the explanation is obvious by the theory of natural selection of slight successive modifications, each profitable in some way to the modified form and often affecting other parts of organization by correlation. These kinds of changes have little or no tendency to modify original patterns or transpose parts. Limb bones may be shortened and widened and become gradually covered in a thick membrane to serve as a fin; some or all bones in a webbed foot may lengthen and the membrane connecting them spread to serve as a wing. Despite such significant modifications, bone framework and the relative connections of the parts would remain unaltered.

The source of homologous limb construction among all mammals is plain if the ancient progenitor, or "archetype," of the whole class had limbs constructed on the existing general pattern. Similarly with insects, if the common ancestor possessed an upper lip, mandibles, and two pairs of maxillae – possibly in simple form – then natural selection accounts for the infinite structural and functional diversity of insect mouths. Nevertheless, the general pattern of an organ can become so obscured as to be ultimately lost. General patterns seem to have been somewhat obscured in the paddles of extinct gigantic sea lizards and in the mouths of certain suctorial crustaceans. Parts may atrophy and eventually disappear, fuse, or double or multiply – all variations known to be within the limits of possibility.

Another interesting branch of this field is the comparison of different parts or organs in the same individual. Most physiologists believe that skull bones are homologous – that is, they correspond in number and relative arrangement – with the elemental parts of certain vertebrae. The anterior and posterior limbs are clearly homologous in members of the vertebrate and articulate classes, and the same rule is observed in the wonderfully complex jaws and legs of crustaceans. It is common knowledge that the essential structure and relative positions of sepals, petals, stamens, and pistils in a flower make sense if they consist of metamorphosed leaves arranged in a spiral. Monstrous plants often provide direct evidence of one organ's ability to transform into another, and organs within embryonic crustaceans, many other animals, or flowers that become very different when fully developed can actually be seen as identical.

These facts are wholly inexplicable with the common notion of creation. Why should the brain be encased in a box composed of so many and so extraordinarily shaped pieces of bone?[6] Why should similar bones have been created in the formation of bat wings and legs, given their to-

6. As Owen remarks, the benefit of yielding skull pieces during mammalian birth does not explain the same construction in birds.

tally different uses? Why should crustaceans with an extremely complex mouth formed of many parts consequently possess fewer legs, and vice versa? Why should the sepals, petals, stamens, and pistils of a flower, with their widely different purposes, all be constructed on the same pattern? The theory of natural selection answers these questions.

Vertebrates have a series of vertebrae-bearing processes and appendages, articulates have a body divided into a series of segments bearing external appendages, and flowering plants have successive whorls of leaves. The characteristic in common is an indefinite repetition of the same part or organ, as observed by Owen with all "low" or little-modified forms. Therefore, the common ancestor of vertebrates probably had many vertebrae, the common ancestor of articulates had many segments, and the common ancestor of flowering plants had many whorls of leaves. As previously discussed, repetitive parts are especially liable to vary in number and structure, so natural selection is likely to seize a subset and adapt its constituents to diverse purposes during a long course of modification. A certain amount of fundamental resemblance in such parts, retained through inheritance, is not surprising, because the total amount of modification is accumulated by slight successive steps.

Among mollusks, although the parts of one species may be homologous with those of another distinct species, rarely have homologies been found *within* a species. This makes sense because even the lowest members of the class do not exhibit nearly so much indefinite repetition of any part as observed in the other major classes of plants and animals.

Naturalists often regard the skull as altered vertebrae, the jaws of crabs as altered legs, and the stamens and pistils of flowers as altered leaves. But as Professor Huxley has remarked, it is probably more accurate in these cases to regard skull and vertebrae, jaws and legs, and so forth as having been altered not one from the other but both from some common element. However, naturalists use this language only metaphorically and do not mean that ancestral organs have actually been modified into skulls or jaws during a long course of descent; yet they can hardly avoid this implication given the strong appearance that such modifications have occurred. *With my theory these terms can be used literally,* explaining, for example, the wonderful case of the jaws of a crab retaining characteristics, probably through inheritance, if they have really been altered during descent, either of true legs or of some simple appendage.

As already mentioned in passing, certain very different organs within an adult individual, serving very different purposes, are identical in the embryo. Also, embryos of distinct animals within a class are often strikingly similar. The best proof of this is an anecdote given by Karl Ernst von Baer, who forgot to label the embryo of some vertebrate and could not then tell whether it was of a mammal, bird, or reptile. The wormlike larvae of moths, flies, beetles, and other insects resemble one another much more than do the adult insects.[7] Embryonic resemblance sometimes lasts until a relatively late developmental stage, so birds within a genus and of closely related genera often resemble one another in their first and second plumage – as in the spotted feathers of the thrush group. Among cats, most species are striped or spotted in lines, and lion cubs are striped. A similar phenomenon sometimes, though rarely, functions in plants. So, for example, the embryonic leaves of furze and the first leaves of phyllode-bearing *Acacias* are both pinnate like the ordinary leaves of most legumes.

Structural elements that resemble one another in the embryos of very different animals belonging to the same class are often not directly relevant to the environment. For example, the peculiar looping course of arteries near the branchial slits in vertebrate embryos cannot be related to similar conditions – the young mammal, nourished in a womb; the bird egg, hatched in a nest; and the frog, spawned under water. There is no more reason to believe in such a relationship than there is to believe that the homologous bones in the human hand, porpoise fin, and bat wing are related to similar environments. No one would suggest that the stripes on a lion cub or the spots on a young blackbird are of any use to these animals or are related to the environments to which they are exposed.

However, the situation is different when an animal is active during any part of its embryonic development and has to provide for itself.

7. Though larvae are active as embryos and have been adapted for special functions.

Regardless of whether the active period comes on early or late in life, larval adaptation to the environment is just as perfect and beautiful as in the adult. These special adaptations sometimes obscure the similarity between larvae or active embryos of related species; examples could be given of larvae of two species, or two groups of species, differing as much or even more from one another than do their adult forms. But in most cases the larvae, though active, obey the rule of common embryonic resemblance. Barnacles offer a good example; even the illustrious Cuvier did not perceive that they are crustaceans, but a glance at the larva unmistakably shows this to be the case. The two main barnacle divisions – pedunculated and sessile – differ greatly in external appearances but have almost indistinguishable larvae.

An embryo generally "rises in organization" during development. I use this expression even though I am aware that it is not possible to clearly define what "higher" or "lower" organization means. But it is reasonable, for example, to describe the butterfly as "higher" than the caterpillar. In some cases, as with certain parasitic crustaceans, the adult is generally considered lower than the larva. Referring again to barnacles, first-stage larvae have three pairs of legs, a very simple single eye, and a prosciformed mouth for copious feeding (they increase greatly in size). Second-stage larvae, corresponding to the butterfly chrysalis, have six pairs of beautifully formed swimming legs, a pair of magnificent compound eyes, and extremely complex antennae. They have a closed and imperfect mouth and cannot feed, their function at this stage being to use their highly developed sense organs to search out and swim to a proper place for attachment and final metamorphosis. This completed, they are fixed for life, their legs made prehensile, their mouth fully functional, but they have no antennae, and their two eyes are reconverted into a minute, single, and very simple eyespot. In this last state, barnacles may be considered more highly or more lowly organized than in the larval state, but in some genera the larva develops either into a hermaphrodite or into what I have called a "complemental male," and here the development is definitely retrograde, because the male is just a sac that lives for a short time and lacks a mouth, stomach, and other important organs except those needed for reproduction.

We are so accustomed to observing structural differences between the embryo and the adult and close similarities between embryos of very different animals within a class that we may be led to consider them necessarily contingent in some way on development. But there is no obvious reason why, for example, the bat's wing or porpoise's fin should not be sketched out with all parts in proper proportion as soon as the structure appears in the embryo. There are entire groups of animals, and certain members of others, in which the embryo is not very different at any period from the adult. Owen has remarked about cuttlefish that "there is no metamorphosis; the cephalopodic character is manifested long before the parts of the embryo are completed," and about spiders that "there is nothing worthy to be called a metamorphosis." Whether insect larvae are adapted to diverse and active habits, or are inactive and fed by parents or placed in the midst of nourishment, almost all pass through a similar wormlike developmental stage. A few insects show no trace of this stage, as documented for aphids by Professor Huxley's admirable drawings of their development.

So how can these elements of embryology be explained? The general, though not universal, structural difference between embryo and adult? The similarity of embryological parts within an individual that ultimately become structurally and functionally very different? The general resemblance of embryos of different species within a class? Embryo structure not being closely related to the environment, except when the embryo is at all active and must provide for itself? The embryo sometimes apparently having a higher organization than the mature adult into which it develops? All of these questions are answered, as follows, by descent with modification.

It is commonly assumed, perhaps because monstrosities often affect embryos at an early stage, that slight variations necessarily appear early. But there is little evidence for this, and it rather points the other way. Breeders of cattle, horses, and various fancy animals cannot positively tell a specimen's ultimate merits and form until sometime after it is born. Similarly for us humans, we cannot always tell whether a child will be tall or short or what its precise features will be. The question is not "At what period of development has a variation been caused?" but "At what period is it fully displayed?" The cause probably acts even before an embryo forms, and the variations may result from the conditions to which either parent's or their ancestors' sexual elements had been exposed. The consequences of an effect at a very early stage – even before embryo formation – may appear late in life, like when a hereditary disease communicated from the reproductive element of one parent appears only in old age, or when the horns of crossbred cattle have been affected by the horn shape of either parent. It is irrelevant to the welfare of a very young animal *exactly when* most of its characteristics develop fully, so long as it remains in the womb, egg, or is nourished and protected by a parent. For example, it would not matter to a bird that best obtained its food by possessing a long beak, whether or not the beak assumed its length early in life, so long as the parents fed it. So I conclude that it may be possible that each of the many successive modifications by which a species has acquired its structure supervened at some not-very-early point in life, as supported by some direct evidence in domestic animals, but in other cases it is possible that each successive modification, or most of them, appeared at a very early period.

As stated in chapter 1, there is evidence that at whatever age a variation first appears in a parent, it appears at the corresponding age in the offspring. Certain variations can *only* appear at corresponding ages; for example, peculiarities in the caterpillar, cocoon, or imago stages of the silkworm or in the horns of almost fully grown cattle. But even more than this, variations that as far as we can tell could appear early *or* late in life tend to appear correspondingly in parent and offspring. This is not always the case, and there are many examples of variations in the broadest sense that supervene at an earlier age in the offspring than in the parent.

If these two principles are accepted, they will explain all the major elements of embryology listed above, but first consider a few analogous cases in domestic varieties. Some authors who have written about dogs maintain that the greyhound and bulldog are really closely related varieties, despite appearing so different, and have probably descended from the same wild stock. I was therefore curious to see how far their puppies differed from each other. Breeders told me that they differed just as much as their parents, and judging by eye this seemed almost completely correct. However, when I actually measured the parents and their six-day-old puppies, I found that the puppies had not acquired their full amount of proportional differences. Similarly, I was told that cart horse and racehorse foals differ as much as the adult animals. This surprised me because the differences between these two breeds were probably caused by selection under domestication. I had careful measurements taken of dams and of three-day-old colts of racehorses and heavy cart horses and found this not to be the case.

The evidence appears conclusive that domestic pigeon breeds have descended from one wild species, so I compared young pigeons of various breeds within twelve hours of hatching. I carefully measured beak proportions, mouth width, nostril and eyelid length, foot size, and leg length in wild stock, pouters, fantails, runts, barbs, dragons, carriers, and tumblers. Now, some of these varieties differ so extraordinarily in beak form and length when mature that they would be ranked in different genera had they been produced in the wild, but when the nestlings were placed in a row, their proportional differences in the above categories were much less than in the adult birds (although most of them could be distinguished from one another). Some characteristic differences, like mouth width, could barely be detected in the young. The one remarkable exception was the young short-faced tumbler, differing in all its proportions from the young rock pigeon and the other breeds almost as much as the adult.

The two principles given above explain these observations with respect to the later embryonic stages of domestic varieties. Breeders select their horses, dogs, and pigeons for breeding when they are nearly adult, indifferent as to whether the desired qualities were acquired early or late in life so long as the adult animal possesses them. The above examples, especially that of the pigeons, show that characteristic differences that give value to a breed and have been accumulated by human selection do not usually appear early in development. However, the short-faced tumbler acquired its proper proportions after twelve hours, proving that this is not a universal rule.

These observations and the above two principles can now be applied to species in the wild. (The second principle has not been proven true, but it is probable.) Consider a genus of birds descended from some one ancestral species, its members having become modified through natural selection according to their diverse habits. The many slight successive variations having supervened at a late age, and having been inherited at a correspondingly late age, *the young of the various species in the genus will manifestly tend to resemble one another much more closely than do the adults,* as with the pigeons. This reasoning can be extended to whole families or even classes. For example, forelimbs that serve as legs in an ancestral species may become, through a long course of modification, hands in one descendant, paddles in another, or wings in a third. But the embryonic forelimbs of the descendants will still resemble one another, being un-modified, because of the above-stated two principles: successive modifi-cations supervene at a late age and are inherited at a correspondingly late age. However, within each *individual* descendant species, the forelimbs will have undergone modifications that are relevant at a late age, so the forelimbs of the embryo and adult will differ greatly. Whatever influence use or disuse may have in modifying an organ, it will affect only the adult that has grown to its full range of activities and must survive indepen-dently, and such effects will be inherited at a correspondingly late age. The young will remain unmodified, or at least less so, by use and disuse.

In certain cases the successive steps of modification may supervene early in development due to unknown causes, and the embryo or young animal would resemble the adult, as with the short-faced tumbler. This form of development applies to certain entire groups, like cuttlefish and spiders, and to a few insects, like the aphids. In these cases, two con-tingencies may cause the young to *not* undergo metamorphosis or re-semble their adult forms from an early age: (1) the young must provide for themselves from an early age, or (2) they must exactly follow adult habits from an early age. With (2) it is indispensible for the survival of the species that the young should be modified in the same way as the parents, in accordance with their similar habits. (A deeper explanation of non-metamorphosing embryos may be required.) If, however, the young profit from habits that are at all different from the adult habits and possess an appropriately dissimilar structure, the active young or larva can easily be rendered completely different from the adults by natural selection. Such differences can become correlated with developmental stages, so the first-stage larva can differ from the second-stage larva, as discussed for barnacles. If the adult is adapted to habits in which organs of sense, locomotion, and so on are useless, the final metamorphosis is termed "retrograde."

All organisms – living and extinct – that have *ever* lived on this earth must be classified together. They are all connected by the finest gradations. The best classification, indeed the only possible one were our collections complete, is genealogical, and common descent is the hidden bond that naturalists have been seeking in the "natural system." This interpretation explains why most naturalists consider embryonic structure more important for classification than adult structure, the embryo being in a "less modified state" and in this sense revealing ancestral structure. If the embryos of two animal groups pass through similar embryonic stages, then they have both descended from a common ancestor: community of embryonic structure reveals community of descent. The rule applies even when the adults differ in structure and habit. For example, as mentioned earlier, barnacles are immediately recognized as crustaceans by inspection of the larvae.

Extinct life forms resemble embryos of their descendants – that is, of existing species. Agassiz believes this to be a law of nature, whereas I only expect it to be proven someday. It can be demonstrated only in cases where the ancestral state – now represented in many embryos – has not been obliterated by successive variations having supervened at early developmental stages. One should also keep in mind that although ancient life forms may resemble embryos of extant forms, the rule may remain unproven for a long time, or forever, because the geological record does not stretch back far enough in time.

So it seems to me that the major elements of embryology, the most important in natural history, are explained by slight modifications appearing in the many descendants of a common ancestor but not early in development, though the modifications may have been *caused* early in development. These modifications become manifest at a late age and are inherited at a corresponding late age in offspring. Embryology rises greatly in interest if the embryo is seen as a picture, perhaps obscured, of the common ancestor to each major class of animals.

Rudimentary, degenerate, and aborted parts and organs, bearing the stamp of uselessness, are very common throughout nature. Examples include rudimentary breasts in the males of many mammals; the "bastard wing" in birds (probably a rudimentary digit); one lobe of the lungs in many snakes; and in other snakes, rudiments of a pelvis and hind limbs. Some cases are very curious; for example, there are teeth in fetal whales that, when grown up, have not a tooth in their heads; unborn calves have teeth in their upper jaws that never so much as cut through the gums; and it has even been stated on good authority that rudimentary teeth can be detected in the beaks of certain embryonic birds. Although it is obvious that wings are formed for flight, many insects have wings that are so reduced as to be utterly useless for flight, often lying under wing cases, firmly soldered together!

The meaning of rudimentary organs is often unmistakable. For example, there are beetles within the same genus, and even within the same species, that closely resemble each other, but one has full wings while the other has just rudiments of membrane. It is impossible to doubt that the rudiments are derived from wings. Rudimentary organs sometimes retain their potentiality, as in the many cases of breasts becoming fully developed in male mammals and secreting milk. Comparably, members of the *Bos* genus normally have udders with four developed and two rudimentary teats, but in domestic cows these two sometimes develop and give milk. Individual plants within the same species sometimes have rudimentary petals and sometimes developed petals. Male flowers of plants with separate sexes often have a rudimentary pistil, but Kölreuter finds that crossing these males with a hermaphroditic species yields a larger rudimentary pistil in the offspring, demonstrating that the rudimentary and perfect pistil are essentially alike.

An organ serving two purposes may become rudimentary or nonfunctional for one purpose – perhaps even the more important one – and remain perfectly functional for the other. For example, in plants, the pistil, consisting of a stigma supported by a style, allows pollen tubes to reach the ovules protected in the ovary at the base. However in some composites, the obviously unfertilizable male florets have a rudimentary pistil, uncrowned by a stigma but retaining a style clothed in hairs for brushing pollen out of surrounding anthers. In certain fish the swim bladder seems to be rudimentary for its primary function – buoyancy – and has been converted into a nascent breathing organ. Other, similar examples could be given.

The rudimentary organs of individuals within a species tend to be variable. Furthermore, among closely related species, the extent to which an organ has become rudimentary sometimes varies, exemplified by the wings of females in certain moth groups. Rudimentary organs are sometimes completely aborted, implying that no trace of an organ exists when analogy suggests it should exist. Otherwise aborted organs sometimes reappear in monstrous individuals, as in the snapdragon's usually absent fifth stamen. The common use and discovery of rudiments is necessary for tracing the homology of a given part among members of the same class, as shown through drawings given by Owen of horse, ox, and rhinoceros leg bones.

It is important that rudimentary organs, like teeth in the upper jaws of whales and ruminants, often appear in the embryo but then disappear entirely. It is also a universal rule that a rudimentary part or organ in an embryo is proportionally larger than in the adult, so at this early stage it is "less rudimentary" or not rudimentary at all. This is why adult rudimentary organs are often described as "retaining the embryonic condition."

Reflecting on these major observations concerning rudimentary organs must lead to astonishment, because the same reasoning power that clearly reveals that most parts and organs are exquisitely adapted for specific purposes also reveals that these rudimentary or degenerate organs are imperfect and useless. Works on natural history generally describe them as created "for the sake of symmetry" or "to complete the scheme of nature." But this is no explanation, merely a restatement of the observation. Would anyone think it sufficient to argue that satellites revolve around the planets elliptically to complete the "scheme of nature," for the sake of symmetry, because planets revolve around the sun elliptically? An eminent physiologist accounts for rudimentary organs by supposing that they excrete excess or harmful material, but can this really be the function of the minute papilla, often representing the pistil in male flowers and formed of merely cellular tissue? Can the rapidly growing embryonic calf really gain anything by excreting precious phosphate minerals to form teeth that are subsequently reabsorbed?

Imperfect nails sometimes appear on the stumps of amputated human fingers, but I would as soon believe that these vestigial nails appear in order to excrete excess keratinous material as I would believe that rudimentary nails on manatee fins form for this same purpose rather than unknown laws of development.

Descent with modification makes the origin of rudimentary organs simple to understand. There are plenty of examples in domesticated organisms: tail stumps, vestigial ears, the reappearance of minute dangling horns in hornless cattle breeds (especially in young animals, according to Youatt), and the state of the flower in cauliflowers. Rudiments of various parts often occur in monsters, but I doubt that any of these cases can illuminate the origin of rudimentary organs in the wild except to show that rudiments can be produced, because I doubt that species in the wild undergo abrupt changes. Disuse has been the main force, leading to the gradual reduction of various organs across successive generations until they have become rudimentary. Examples include the eyes of animals inhabiting dark caverns and the wings of birds inhabiting oceanic islands, the owners having been seldom forced to take flight and ultimately losing the ability. But to reiterate, an organ useful in one environment may be harmful in another, like the wings of beetles living on small exposed islands. In these cases natural selection slowly and continually reduces the organ until it is rendered rudimentary, and harmless.

Any change in function that can be effected by small steps is within the power of natural selection, so if the environment changes and an organ becomes useless or harmful for one purpose, it can easily be modified for another purpose. Alternatively, an organ can be retained for only one of its previous functions. An organ rendered useless is likely to be variable, because its variations can no longer be checked by natural selection. Disuse or selection generally reduces an organ in the adult, when the organism reaches full functionality; the principle of inheritance at corresponding ages will reproduce the reduced organ at the same age in offspring and, consequently, will rarely affect it in the embryo. This explains the relatively large size of rudimentary organs in embryos and their relatively small size in adults. If each step of reduction is inherited early in development instead of at a corresponding age, as there is good reason to deem possible, the rudimentary part would eventually be wholly lost. Also, the principle of economy described in chapter 5, by which materials forming a useless part or structure are conserved as much as possible, probably contributes to the obliteration of rudimentary organs.

So rudimentary organs persist because every part that has long existed tends to be passed on by inheritance. Supplementing the genealogical interpretation of classification, this explains why systematists find rudimentary parts as useful, or sometimes even more useful, than parts of high physiological importance. Rudimentary organs are like silent letters in a word, retained in the spelling and a clue to its origins. The existence of rudimentary, imperfect, useless, or aborted organs definitely presents a strange difficulty for the common doctrine of creation, but not for descent with modification, which may even have anticipated the phenomenon, accounted for by inheritance.

The organization of all organisms throughout all time into groups subordinate to groups, the complex and radiating relationships uniting all living and extinct creatures into one grand system, the rules followed and difficulties encountered by naturalists in classification, the values of various types of characteristics in classification, the wide opposition in value of analogical and real characteristics, and other such rules all follow naturally from the view that co-classified organisms share a common ancestor, together with their modification through natural selection, with its contingencies of extinction and divergence. In evaluating this interpretation, note that the element of descent has been universally applied in grouping together the sexes, ages, and acknowledged varieties of the same species, however different they may be in structure. Extending this element of descent – *the only known cause of similarity in organisms* – illuminates the meaning of the natural system: it is *genealogical* in its attempted arrangement, with the grades of acquired difference marked as varieties, species, genera, families, orders, and classes.

Descent with modification also renders intelligible the main elements of morphology, whether considering the similarity of pattern in homologous organs of different species within a class or the similarity of pattern in homologous parts within each individual plant and animal.

The rule that slight successive variations do not necessarily or generally supervene early in development explains the major elements of embryology: the resemblance of homologous parts in the embryo that become very different in structure and function when fully developed, and the resemblance of homologous parts or organs in different species within a class, though fitted for very different purposes.

The observations considered in this chapter seem to proclaim so clearly that the countless species, genera, and families populating the earth – each within its own group – have all descended with modification from common ancestors that I would adopt this interpretation even if it were unsupported by other observations or arguments.

14

SUMMARY AND CONCLUSION

AS THIS ENTIRE BOOK IS ONE LONG ARGUMENT, IT MAY BE convenient to briefly review the main observations and inferences. I do not deny that many and serious objections can be advanced against the theory of descent with modification by means of natural selection, and I have endeavored to grant them full force. At first it seems very difficult to believe that complex organs and instincts are perfected by the accumulation of countless minor variations and not by means superior to, though analogous with, human reason. This difficulty may seem beyond our imagination, but it cannot be real if we accept the following: (1) Gradations leading to the perfection of any organ or instinct, each one good for its individual possessor, *do* exist or *could have* existed. (2) All organs and instincts are ever so slightly variable. (3) There is a struggle for existence during which each profitable deviation of structure or instinct is preserved. I do not think the validity of these propositions can be disputed.

It is extremely difficult to even conjecture by what gradations many structures have been perfected, especially in fragmenting and failing groups. But many gradations do exist in nature, as reflected by the saying *natura non facit saltum*, so we should be very cautious in asserting that any organ, instinct, or whole organism could not have been produced by many graded steps. Some cases do pose a special challenge to the theory of natural selection, one of the most curious being the sterile worker castes of some ant species, but I have tried to show how this can be explained.

The nearly universal sterility of crosses between different species is a remarkable contrast to the near universal fertility of crosses between varieties. The summary at the end of chapter 8 conclusively shows that interspecies sterility is incidental to constitutional differences of the reproductive systems between the two species; it is no more a special endowment than the incapacity of certain plants to be grafted together. This conclusion is supported by the vast difference in the results of the two halves of a reciprocal cross.

The fertility of intercrossed varieties and their mongrel offspring is not an absolute rule, but its generality should be expected, given that the constitutions and reproductive systems of varieties have not been profoundly modified. Moreover, most of the varieties that have been studied are domesticated, and as domestication tends to eliminate sterility, it should not be expected to also produce sterility.

Species sterility is very different from hybrid sterility: when individuals of different species are crossed, their respective reproductive organs are in a perfect condition, whereas the reproductive organs of hybrids are somewhat functionally impotent. All kinds of organisms experience reduced fertility if their constitutions are disturbed by changes to the environment. As hybrids are compounded from two distinct organisms, it is not surprising that their constitutions are disturbed. This parallelism is supported by another: all organisms experience increased fertility and vigor from small environmental changes, and varieties have small differences between them, so their offspring gain fertility and vigor from the cross. Considerable changes to the environment and crosses between highly modified forms decrease fertility, whereas smaller changes to the environment and crosses between slightly modified forms increase fertility.

Geographic distribution presents significant difficulties for the theory of descent with modification. All individuals of a species and all species of a genus, or even higher group, must have descended from common ancestors. Therefore, regardless of current physical separation or isolation, they must have moved from one region to the others over the course of successive generations. It is often impossible to guess how this may have happened. The wide diffusion of certain species should not be stressed too much, because some species have maintained a constant form over enormously long periods, allowing for long migrations by many means. An interrupted range may often indicate extinction of the species in the intermediate regions. We are still very ignorant of the full extent of the changes in climate and geology that have recently affected earth, and such changes have obviously greatly facilitated migration. As an example I tried to show the potent influence of the glacial period on the distribution of certain species. We are also very ignorant of accidental means of dispersal. Distinct species from within one genus occupying very distant and isolated regions have come to differ by the necessarily slow process of modification, and all means of migration are possible on such large timescales.

The theory of natural selection predicts an interminable number of intermediate forms that link together all the species in a given group by fine gradations – as fine as those linking varieties. So why aren't these linking forms all around us? Why aren't all organisms blended together in an inextricable chaos? Recall that with a few rare exceptions, there are no directly linking forms between existing species, *only between each species and some extinct ancestor.* Even in a big area that has remained continuous for a long time, with a climate and environment that changes gradually from a region occupied by one species to a region occupied by another, there is no reason to expect intermediate varieties in the intermediate zone, because only a few species are undergoing change at any given time, and change is slow. I also showed that intermediate varieties *do* initially exist but are subject to extermination by the related forms on either side, which exist in greater numbers and are therefore modified and improved more quickly.

If an infinite number of connecting links between the living and extinct inhabitants of the world have been exterminated at each successive geological age, then why isn't every geological formation suffused with them? Why doesn't every fossil collection afford clear evidence of the gradation and mutation of life? There is no such evidence, and this objection is the most obvious and most forcible of many objections against my theory. Why do whole groups of related species appear suddenly in geological strata? (Certainly this is often a false appearance of simultaneity.) Why aren't there great piles of strata beneath the Silurian system stored with the remains of the fossilized ancestors of Silurian species? My theory positively predicts that such strata must somewhere have been deposited during these ancient and wholly unknown periods of earth's history.

I can address these questions and objections only on the assumption that the geological record is far more imperfect than most geologists believe. However, it cannot be objected that time has been insufficient for any amount of organic change; its lapse has been so great as to be utterly incomprehensible to human intellect. All specimens in all museums amount to nothing in comparison to the countless generations of countless species that have existed. It is impossible to recognize one species as the ancestor of any other without possession of the many intermediate links between the ancestral and descendant states, and these many links cannot be discovered owing to the imperfection of the geological record. Many existing difficult-to-categorize forms are probably varieties, but who will pretend that in the future *so* many fossil links will be discovered that naturalists will be able to decide, based on the commonly accepted notions, whether or not they are varieties? So long as most of the links between any two species remain unknown, if any one link or intermediate is discovered it will simply be categorized as another distinct species. Only a small part of the world has been geologically explored, and only organisms of certain classes can be fossilized (at least, in any great number). Widely ranging species vary the most, and varieties are often local at first; both conditions render the discovery of intermediates less likely. Local varieties do not spread into other distant regions until they are considerably modified and improved, and when they *do* spread, they will appear in geological formations as having been suddenly created and will simply be classed as new species. Most formations accumulate intermittently, and their duration, I believe, is probably shorter than the average duration of specific forms. Successive formations are separated from each other by enormous blanks of time. Fossil-rich formations thick enough to resist future degradation can be accumulated only where copious loads of sediment are deposited on the subsiding bed of the sea. During alternate periods of elevation, or when the seabed remains stationary, the record remains blank; during elevation there will probably be more variability in the forms of life, and during subsidence, more extinction.

Concerning the absence of fossils beneath the lowest Silurian strata, I can only recur to the hypothesis given in chapter 9. Everyone agrees that the fossil record is imperfect, but very few agree that it is as imperfect as I require. At long enough intervals of time, geology plainly proclaims that all species change, and they change in the time required by my theory: slowly and gradually. This is manifest in the fossil remains of consecutive formations being more closely related to each other than are fossils from formations distant in time.

Such is the sum of the main objections and difficulties that may justly be brought against my theory, and I have briefly reviewed the answers and explanations that can be given to them. I have felt these difficulties far too heavily during many years to doubt their weight, but it deserves special notice that the more important objections concern issues of which we are ignorant. Even the very extent of our ignorance is unknown. We do not know all the possible transitional gradations between simple and perfect organs. We do not know all the means of distribution during the long lapse of years. We do not know how imperfect the geological record is. Grave as these objections are, in my judgment they do not overthrow the theory of descent with modification.

Now for the other side of the argument. Under domestication there is much variability. This seems to result mainly from the reproductive system's susceptibility to environmental changes such that it fails to reproduce offspring exactly like the parent. Variability is governed by many complex rules, including correlated growth, use and disuse, and the direct influence of the physical environment. It is difficult to ascertain how much domesticated organisms have changed though safe to infer that they have changed a lot and that modifications can be inherited for long periods. So long as the environment remains constant, there is reason to believe that a modification already inherited for many generations will continue to be inherited almost indefinitely. However, there is evidence that once variability starts, it does not wholly cease; new varieties are still occasionally produced by even the most anciently domesticated organisms.

Humans do not actually generate variability; they only unintentionally expose organisms to new environments where nature acts on the organization and causes it. But humans can and do *select* these nature-supplied variations, accumulating them in any desired direction and thereby adapting animals and plants for human benefit or pleasure. This may be done methodically, or it may be done unconsciously by the preservation of individuals that are most useful at some given time without any thought to altering the breed. It is certain that a breed can be significantly altered by the selection in each successive generation of individual differences that are so slight as to be inappreciable to the untrained eye. This process has produced the most distinct and useful domestic breeds. Domesticated breeds produced by humans are to a large extent like species, as shown by the inextricable uncertainties over whether many of them are varieties or original species.

There is no obvious reason why mechanisms so efficient under domestication should not function in nature. The powerful and ever-acting means of selection is the preservation of favored individuals and varieties during a constant struggle for existence. The struggle for existence follows inevitably from the high geometrical ratio of increase common to all organisms. This high rate of proliferation is provable by calculation, by the effects of a succession of peculiar seasons, and by the results of naturalization, as explained in chapter 3. More individuals are born than can possibly survive. A grain in the balance determines which individual lives and which dies, which variety or species proliferates and which declines or becomes extinct.

The closest competition, in all respects, is between individuals of the same species. The struggle is severest between them. Almost equally severe is the struggle between varieties of the same species, followed by the struggle between species of the same genus. The struggle is often very severe even between organisms that are far apart on the scale of nature. A slight advantage to one organism at any age or season over its competitors, or better adaptation to the environment – however slight – tips the balance.

In animals with separate sexes there is often a struggle between the males for the possession of females. The most vigorous, or the most successful in their struggle with the environment, generally leave the most progeny. But success often depends on special weapons, means of defense, or charms of the male. The slightest advantage leads to victory.

Geology clearly reveals the great physical changes endured by each land, so the variability of organisms in the wild should be expected, given that they generally vary under the changed conditions of domestication. And if there is any variability in the wild, it would be an unaccountable fact if natural selection did *not* act. An assertion that is often made but cannot be proven is that the total amount of available variation in the wild is limited. Though humans act on external characteristics alone, and often capriciously, they can produce great results within short times by adding up mere individual differences. Moreover, everyone agrees to the existence of at least individual differences among species in the wild, and besides these, all naturalists agree to the existence of varieties that are sufficiently distinct to warrant record in works of classification. No clear distinction can be drawn between individual differences and minor varieties, or between well-defined varieties and sub-species, or between sub-species and species. Note how naturalists rank many representative European and North American forms differently.

So if there is variability in the wild, and a powerful agent always ready to act and select, why doubt that variations that are at all useful to organisms in their extremely complex relationships of life are preserved, accumulated, and inherited? If humans can patiently select variations useful to them, why should nature fail to select variations useful to living things in a changing environment? What could limit this great power, acting during long ages and firmly scrutinizing the whole constitution, structure, and habits of each creature, favoring the good and rejecting the bad? I can see no limit to this power in slowly and beautifully adapting each form to the most complex relationships of life. Even if nothing further is added, the theory of natural selection in itself seems probable to me.

The view that species are simply well-defined and permanent varieties and that each species begins as a variety explains why it is not possible to draw a demarcation between species – commonly supposed to have been produced by special acts of creation – and varieties – acknowledged to have been produced by secondary causes. This same view explains why in a region where many species of a genus have been produced there are also many varieties of these species; as a general rule, wherever many species have been generated, they should still be generated. This is expected if varieties are incipient species. Furthermore, species of large genera, which afford a greater number of incipient species, retain the character of varieties because they differ less from one another than do species of small genera. The closely related species of large genera also have restricted ranges, and they cluster in little groups around other species as varieties do. These are strange relationships if species are thought to have been independently created but are intelligible if all species exist first as varieties.

Each species tends to proliferate inordinately, and the modified descendants of each species are all the better enabled to proliferate as they diversify in habits and structure so that they can seize many and very different niches. There is a consequent tendency for natural selection to preserve the most divergent descendants of any given species. Thus, over time, the small differences typical of varieties tend to become augmented into the greater differences typical of species. New and improved varieties inevitably supplant and exterminate older, less improved, and intermediate varieties, rendering species mostly defined and distinct objects. Dominant species of large groups tend to generate new and dominant forms, so each large group tends to become larger and more divergent. But *all* the groups cannot increase in size this way, because the earth cannot hold them, so the dominant groups beat the less dominant. This tendency, together with the almost inevitable contingency of extinction, explains the arrangement of all forms of life in groups subordinate to groups, all within a few great classes, which we now see all around us and which has prevailed throughout all time. This grand feature of the grouping of all organisms is completely inexplicable by the theory of creation.

Natural selection acts solely by accumulating slight, successive, and favorable variations in short and slow steps, so it cannot produce great and sudden modifications. *Natura no facit saltum* becomes more strictly correct with every fresh piece of knowledge. It is made intelligible by this theory, which explains why nature is prodigal in variety but niggardly in innovation. But no one can explain why this should be a rule of nature if each species has been independently created.

Many other observations are also explicable through this theory of descent with modification. How strange it would be if a bird in the form of a woodpecker had been created to prey on ground insects. How strange it would be if upland geese, which rarely or never swim, had been created with webbed feet. How strange it would be if a thrush had been created to dive and feed on aquatic insects. How strange it would be if a petrel had been created with habits and structure befitting an auk or grebe. How strange an endless number of other cases would be if each species has been independently created! But these observations cease to be strange – and can perhaps even be anticipated – if one recognizes that species constantly tend to proliferate, with natural selection always ready to adapt the slowly varying descendants of each to any unoccupied or poorly occupied niche.

Natural selection acts through competition, so it can only adapt the inhabitants of a region relative to one another. Consequently, it is not surprising that the inhabitants of one region can sometimes be beaten and supplanted by invasive organisms from another region. (The common notion holds that each has been specially created and adapted for its own region.) It is therefore unsurprising that not all natural structures are absolutely perfect and that some of them are abhorrent to our conception of "fitness"; for example, the bee's sting causes its own death, drones are produced in vast numbers for one act and then slaughtered by their sterile sisters, pollen is wasted astonishingly by fir trees, the queen bee instinctively hates her own fertile daughters, and ichneumonidae feed within the living bodies of caterpillars. In fact, it is surprising that more cases of want of absolute perfection have not been observed.

As far as we can tell, the complex and poorly understood rules governing variation are the same as those governing the generation of so-called species. In both cases the physical environment seems to induce only a minor direct effect (although when some varieties enter any zone, they occasionally assume some characteristics of species in that zone). In both varieties and species, use and disuse seem to produce some effect. This conclusion is difficult to resist, given, for example, the loggerheaded duck, with its wings in nearly the same condition as those of the domestic duck and useless for flight; or the burrowing tucutucu, which is occasionally blind; and certain moles that are habitually blind and have eyes covered with skin; or the blind animals inhabiting the dark caves of America and Europe. In both varieties and species, correlated growth seems very important so that when one part is modified, other parts also necessarily change. In both varieties and species, reversions to long-lost characteristics occur. The occasional appearance of stripes on the shoulder and legs among members of the horse genus is wholly inexplicable by the theory of creation! But it is simply explained if these species have descended from a striped common ancestor, in the same way that domestic pigeon breeds have descended from the blue and barred rock pigeon.

If each species has been independently created, why should specific characteristics vary more than generic characteristics? For example, why should a flower's color be more likely to vary if other members of its genus – supposedly created independently – have different-colored flowers than if they have the same-colored flowers? This observation is understandable if species are just well-defined varieties with characteristics become fairly permanent; they have come to vary in certain characteristics since branching off from a common ancestor, thereby becoming specifically distinct. These same characteristics are more likely to *still* vary than generic characteristics, which have been inherited without change for long periods. The theory of creation cannot explain why an unusually developed, and consequently very important, part in one species of a genus should be especially liable to variation. But according to my view, this part has undergone an unusual amount of variability and modification since branching off from the common ancestor of the genus and might therefore still be generally variable. A part can nevertheless be unusually developed – like the wing of a bat – and be no more variable than any other structure if it is common to many subordinate forms (i.e., if it has been inherited for a long time). In this case the structure has been rendered constant by continuous natural selection.

Marvelous though some of them are, instincts are no more difficult to explain through the natural selection of successive, small, but profitable modifications than bodily structures. This is why different animals within the same class possess graded instincts. I attempted to show how the principle of graded instincts illuminates the architectural powers of honeybees. Habit sometimes modifies instincts, but this is not essential, as illustrated by neuter insects, which cannot leave progeny to inherit the effects of habit. Related species possess nearly the same instincts in considerably different environments, because all species within a genus

descend from a common ancestor and inherit much in common. For example, the South American thrush lines its nest with mud, like the British thrush. And, instincts having been slowly acquired through natural
selection, it is unsurprising that some are not perfect and prone to error,
or that many of them cause other animals to suffer.

If species are well-defined and permanent varieties, it is understandable that the offspring of crosses between species follow the same
complex rules in the extent and nature of resemblance to parents as the
offspring of crosses between varieties. This would be strange if species
have been independently created and varieties have been produced by
secondary causes.

If it is accepted that the fossil record is extremely imperfect, then
such facts as it provides support the theory of descent with modification. New species come on the stage slowly and at successive intervals,
with the amount of change, after equal intervals of time, differing widely
for different groups. The extinction of species and of entire groups of
species figures conspicuously in the history of life and follows almost
inevitably from natural selection: old forms are supplanted by new and
improved forms. Extinct forms do not reappear once the chain of generation breaks. Dominant forms diffuse gradually and their descendants become modified slowly, but after long intervals of time they appear to have
changed simultaneously across the earth. That fossils are intermediate to
those found in formations that are above and below is readily explained
by their intermediate position in the chain of descent. The grand fact
that all extinct organisms belong to the same system as extant organisms – falling into the same or intermediate groups – follows from the
shared common ancestry of extinct and extant forms. The groups that
descend from some common ancestor generally diverge in characteristics, so the ancestor and its *early* descendants often have characteristics
intermediate to those of its descendant groups; this is why the more ancient a fossil, the more likely it is to stand intermediate between related
existing groups. Recent forms are generally considered "higher," in some
vague sense, than extinct forms, and they are insofar as improved forms
supplant older and less improved organisms in the struggle for existence.
Lastly, the endurance of related forms on a continent (e.g., marsupials

in Australia and the Edentata in America) is intelligible because extinct and extant forms will naturally be related by descent within a confined region.

The theory of descent with modification also renders the major observations of geographic distribution understandable if over long periods of time there has been significant migration across the earth, resulting from climatic and geographic changes and the many poorly understood means of dispersal. There is a striking parallelism in the distribution of organisms throughout space and their geological succession throughout time, for in both cases organisms are linked by the bond of generation and the mode of modification is the same. Every traveler must be struck by the wonderful observation that on one continent, under the most diverse conditions – heat and cold, mountain and lowland, desert and marsh – most inhabitants within each major class are clearly related. This is fully comprehensible because they are generally descendants of the same ancestors and early colonists. The same principle of migration, combined in most cases with modification and aided by the glacial period, explains why certain plants are identical and many others are related on very distant mountains, in different climates; it also explains the close relationship between some inhabitants of the sea in the northern and southern temperate zones, though separated by the whole intertropical strip of the oceans. Although two regions may have the same physical environment, their inhabitants may be very different if they have been completely separated for a long time. The relationships between organisms are the most important kind of relationship, and as the two regions received colonists from some third source or each other at various times and in unequal proportions, the course of modification in the two areas will have inevitably been different.

This concept of migration coupled to subsequent modification explains why oceanic islands are inhabited by few but mostly unique species. It also shows why animals that are incapable of crossing wide tracts of ocean, such as frogs and terrestrial mammals, are absent from oceanic islands while new and unique species of bats, which can cross the ocean, are often present on islands far distant from any continent. Such observations are completely inexplicable on the theory of independent acts of creation.

The existence of closely related species in any two areas implies that the same ancestors once inhabited both, and it is often true that whenever many closely related species inhabit two areas, some identical species persist as common to both. Wherever many closely related yet distinct species occur, many varieties of these species also occur. It is a general rule that inhabitants of an area are related to the inhabitants of the nearest potential source of immigrants. This is observed in the striking relationship of nearly all the plants and animals on the Galápagos Archipelago, Juan Fernandez, and other American islands to those of the American mainland, and of those on the Cape de Verde Archipelago and other African islands to those of the African mainland. It must be admitted that these observations receive no explanation on the theory of creation.

The fact that all past and present life constitutes a grand natural system, with group subordinate to group and with extinct groups often falling in between extant groups, is intelligible through the theory of natural selection, with its contingencies of extinction and divergence. These same principles explain why the relationships among species and genera within each class are so complex and circuitous. They explain why some characteristics are useful for classification while others are not; why analogical characteristics, although of supreme importance to the organism, are of almost no importance to classification; why rudimentary characteristics, although of no importance to the organism, are of high importance to classification; and why embryological characteristics are the most valuable of all. The true relationships among all organisms are due to inheritance – that is, community of descent. The natural system is a genealogical arrangement in which the lines of descent are discovered through the most permanent characteristics, however slight their vital importance may be.

The identical bone framework in the human hand, bat wing, porpoise fin, and horse leg; the identical number of vertebrae in the giraffe's neck and the elephant's neck; and countless similar facts immediately explain themselves through the theory of descent with slow and minor successive modifications. The similarity of pattern in the wing and leg of a bat, although used for such different purposes; or in the jaws and legs of the crab; or the petals, stamens, and pistils of the flower are likewise intelligible through the gradual modification of parts or organs, which were alike in the early ancestor of each class. Successive variations do not always supervene at early developmental stages and are inherited to become manifest at a corresponding age in offspring, which is why the embryos of mammals, birds, reptiles, and fish are so alike yet their adult forms are so unalike. It is no longer mysterious that embryos of air-breathing mammals and birds have gill slits and arteries running in loops like those in a fish, which breathes air dissolved in water through fully developed gills.

An organ rendered useless by new habits or a changed environment tends to become reduced through disuse, sometimes aided by natural selection. This renders the meaning of rudimentary organs understandable: disuse and selection generally act on fully developed creatures that are wholly invested in the struggle for existence and so have little power to act on an organ in early life. For example, the calf has inherited teeth that never cut through the gums of the upper jaw from an early ancestor with well-developed teeth. The teeth of the fully developed animal were reduced during successive generations by disuse or by natural selection fitting the tongue and palate to browse without their aid. The teeth of the calf were left unaltered by disuse or selection, and they still have them by the principle of inheritance at corresponding ages. If each organism and each organ has been separately and specially created, it is completely inexplicable that the teeth of the embryonic calf, the shriveled wings beneath the fused wing covers of some beetles, and other such parts should so often bear the stamp of inutility. It is as if Nature had taken pains to reveal through rudimentary organs and homologous structures her scheme for modification, which we seem willful not to understand.

I have now reviewed the major observations and considerations that have thoroughly convinced me that species have changed, and are still slowly changing, by the preservation and accumulation of minor favorable variations. Why, then, have all the most eminent living naturalists and geologists rejected this view of the mutability of species? It cannot be asserted that organisms in the wild do not vary. It cannot be proven that the total amount of variation across the ages is limited. No clear distinction can be made between species and well-defined varieties, and it cannot be maintained that crossed species are invariably sterile, that crossed varieties are invariably fertile, and that sterility is a special endowment and sign of creation. The belief that species are immutable was almost unavoidable so long as the world was thought to have existed for a short time. Now that we have some idea of the lapse of time, we are too apt to assume, without proof, that the geological record should be so perfect that it would afford clear evidence of the mutation of species if they had undergone mutation.

But the main cause of our natural unwillingness to accept that one species has given birth to another is that we are always slow to admit any great change if the intermediate steps are not visible. This unwillingness is the same as that felt by so many geologists when Lyell first asserted that long lines of inland cliffs and great valleys have been formed by the slow lapping of coastal waves. The mind cannot possibly grasp "one hundred million years." It cannot add up and perceive the full effects of many minor variations accumulated for an almost infinite number of generations.

I am fully convinced of the truth of the concepts given in this book, but I do not expect to convince experienced naturalists whose minds are stocked with a multitude of facts all considered, for many years, from a point of view directly opposite to mine. It is so easy to hide our ignorance under such expressions as "the plan of creation," "unity of design," and so on, and it is so easy to think that a restatement of fact is an explanation. Anyone with a disposition to lay greater weight on unexplained difficulties than explained observations will reject my theory. A few naturalists endowed with flexible minds and already doubting that species are immutable may be influenced by this book. But I look with confidence to the future when young and rising naturalists will consider both sides of the question with impartiality. Anyone who comes to believe that species are mutable will do good service by expressing his conviction. This is the only way to remove the load of prejudice overwhelming our subject.

Several eminent naturalists have recently published their belief that many supposed species in each genus are not *real* species but that others *are*, having been independently created. This is a strange conclusion. They admit that many forms – which until recently they thought were special creations and which the majority of naturalists still think are special creations – were produced by variation, but they refuse to extend this concept to other forms that are only very slightly different. They do not pretend they can define, or even conjecture, which forms were created and which were produced by secondary causes. They admit variation as a true cause in one case yet arbitrarily reject it in another without defining any distinction in the two cases. One day this will be given as a curious illustration of the blindness of preconceived opinion.

These authors seem no more startled at miraculous acts of creation than ordinary birth. But do they really believe that on countless occasions in earth's history, elemental atoms have been commanded suddenly to flash into living tissues? Do they believe that each supposed act of creation generated one individual, or many? Were the infinitely numerous plants and animals created as seeds and eggs, or as fully grown? Were mammals created with the false marks of nourishment from the mother's womb? Naturalists very properly demand a full explanation of every difficulty from those who believe in the mutability of species, but on their own side, in reverent silence, they ignore the entire problem of how species first arise.

How far can the doctrine of modification be extended? The question is difficult to answer because the more distinct the two forms under consideration, the more the force of the arguments falls away. But some of the strongest arguments extend very far. All the members of a given class can be connected together by chains of relationships, and all can be classified the same way: in groups subordinate to groups. Fossil remains sometimes fill up large intervals between existing orders. Rudimentary organs clearly indicate that an early ancestor possessed the organ in fully functional form, and this sometimes necessarily implies an enormous amount of modification in the descendants. Throughout whole classes, various structures form on the same pattern, and the species closely resemble one another as embryos. So I cannot doubt that the theory of descent with modification embraces all the members belonging to a given class. I believe that animals have descended from at most only four or five ancestors, and plants from the same number or fewer.

Analogy would lead me one step further to conclude that all plants and animals descend from some one prototype. But analogy may be a deceitful guide. Nevertheless, all living things share much in common: their chemical composition, germ cells, cellular structure, and the rules governing their reproduction and development. This is manifested even in such minor phenomena as the same poison affecting plants and animals similarly, or the poison of the gall fly producing monstrous growths on both the wild rose and the oak. And so it is probable that all organisms that have ever lived on this earth have descended from one primordial form, into which life was first breathed.

If the concepts of this book on the origin of species are generally accepted, we can dimly foresee a revolution in science. Systematists will still pursue their labors but will no longer be incessantly haunted by shadowy doubt as to whether this form or that form is "in essence" a species. (From my own experience, I am sure this will prove no slight relief.) Endless disputes over whether or not some fifty species of British brambles are "true species" will cease. Not that it will be easy, but systematists will have only to decide whether a form is sufficiently constant and distinct to be definable, and if it is, whether the differences are important enough to warrant a specific name. The second part will become far more essential a consideration than it is now, because even minor differences between two forms unconnected by intermediates are considered by most naturalists as sufficient to define both forms as species. But we are now compelled to acknowledge that the only distinction between species and well-defined varieties is that varieties are known or believed to be connected by existing intermediates, whereas species *used to be* connected the same way. It is possible that forms currently considered just varieties will be thought worthy of specific names, as with the primrose and cowslip. In short, "species" must be treated the way some naturalists treat "genera" – as merely artificial combinations made for convenience. This may not be a cheering prospect, but at least we will be free of the vain search for an undiscoverable essence to the term "species."

The other, more general components of natural history will greatly increase in interest. Terms used by naturalists like "affinity," "relationship," "community of type," "paternity," "morphology," "analogical characteristics," and "rudimentary" will no longer be abstract metaphors. They will have clear meaning. When the organism is no longer seen as something entirely beyond comprehension, when every product of nature is regarded as having had a *history*, when every complex structure and instinct is contemplated as being the summation of many changes, each useful to the possessor (similarly to the way any great mechanical device is viewed as the summation of the labor, experience, and even blunders of many workers), natural history will become so much more interesting!

A grand and almost untrodden field of inquiry will open to discover the causes of variation and correlated growth, the effects of use and disuse, the direct influence of the environment, and other factors. The study of domestication will become immensely more valuable. A new variety raised by humans will be far more important and interesting to study than the addition of yet another species to the infinitude of already recorded species. Classification will become genealogical (as far as it can be made so) and will then truly give what may be called "the plan of creation." The rules of classification will no doubt become simpler once a definite structure is in place. We possess no pedigrees or heraldic guides, so the many diverging lines of descent in natural genealogies must be discovered and traced by characteristics of *any* heritable kind. Rudimentary organs will speak infallibly to the nature of long-lost structures. Species called "aberrant," or more fancifully "living fossils," will help us contemplate ancient life forms. Embryology will reveal the structures of the prototypes of each major class (albeit in somewhat obscured form).

Once we are assured that all individuals of a species and all closely related species of most genera have descended relatively recently from one parent and migrated from some one birthplace, and when we better know the many means of migration, the migrations of earth's former inhabitants will be traceable, given geology's illumination of former changes in climate and the level of the land. Even now some light can be thrown on ancient geographic distribution by comparing the different inhabitants of the sea on opposite sides of a continent, and by the nature of the various inhabitants of that continent relative to their apparent means of immigration.

The great science of geology loses power from the extreme imperfection of the record. The crust of the earth, with its embedded remains, is not a well-filled museum, but rather a poor collection made haphazardly and at large intervals. The accumulation of each fossil formation will be recognized as having depended on an unusual concurrence of circumstances, and the blank intervals between successive stages as representing vast time. But the duration of these intervals will be gauged with some accuracy by the comparison of preceding and succeeding organic forms. Caution is needed in attempting to correlate two formations as strictly contemporaneous that include a few identical species based on the general succession of their life forms. Species are produced by slowly acting causes, not by miraculous acts of creation or catastrophes, and the most important cause of organic change is the set of relationships among organisms, which is almost independent of altered – and perhaps suddenly altered – physical environment. The improvement of one form entails improvement or extinction of others. Consequently, the amount of organic change in the fossil record probably serves as a fair measure of the actual lapse of time. However, the accuracy of this measure should not be overrated, because a number of species within a unit of elapsed time might remain unchanged, and some might migrate into new regions, encounter new associates, and become modified. During the early periods of earth's history, when the forms of life were probably fewer and simpler, the rate of change was probably slower. At the first dawn of life, when very few forms of the simplest structure existed, the rate of change may have been extremely slow. The whole of human history as it is now known may be of a length incomprehensible to us, but it will hereafter be recognized as a fragment of time in comparison to the ages that have elapsed since the first creature – the ancestor of countless extinct and living descendants – was created.[1]

In the distant future I see open fields for far more important research. Psychology will find a new foundation based on the necessary acquisition of each mental capacity by gradation. Human origins and human history will be illuminated.

Authors of the highest eminence seem to be fully satisfied with the notion that each species has been independently created. To my mind, it accords better with what we know of the laws impressed on matter by the Creator that the generation and extinction of species should result from secondary causes, like those determining the birth and death of individuals. When I see all living things not as special creations but as

---

1. [Current estimates place the start of life on earth at about 3.8 billion years ago, and possibly earlier, based on microfossils and corresponding geological data. The earth formed about 4.6 billion years ago, a value derived from geology and astrophysics. By comparison, the Silurian period often referred to by Darwin represents a time only about 400 *million* years ago. – D.D.]

lineal descendants of some few beings that lived long before the first bed of the Silurian system was deposited, I see them ennobled. Judging from the past, we may safely infer that not one living species will transmit its unaltered likeness to a distant future, and very few will transmit progeny of any kind to the far distant future. The way in which all organisms are grouped shows that most species of each genus, and *all* the species of many genera, leave no descendants. They become extinct. We can take a prophetic glance into the future and foretell that common and widely ranging species of large and dominant groups will ultimately prevail and generate new and dominant species. As all life forms are descendants of those that lived long before the Silurian period, we may feel certain that the ordinary succession by generation has never once been broken, and no cataclysm has desolated the whole world. Thus we may anticipate with confidence a future of equally incomprehensible length. And as natural selection works solely through and for the good of each being, all bodily and mental endowments will tend to progress to perfection.

It is interesting to contemplate an entangled bank, clothed with many plants of many kinds, with birds singing on the bushes, with various insects flitting about, and with worms crawling through the damp earth, and to reflect that these elaborately constructed forms, so different from each other, and dependent on each other in so complex a manner, have all been produced by laws acting around us. These laws, taken in the largest sense, being growth with reproduction; inheritance that is almost implied by reproduction; variability from the indirect and direct action of the external conditions of life, and from use and disuse; a ratio of increase so high as to lead to a struggle for life, and as a consequence to natural selection, entailing divergence of character and the extinction of less-improved forms. Thus, from the war of nature, from famine and death, the most exalted object we are capable of conceiving – namely, the production of the higher animals – directly follows. There is grandeur in this view of life, with its several powers, having been originally breathed into a few forms or into one; and that, while this planet has gone cycling on according to the fixed law of gravity, from so simple a beginning endless forms most beautiful and most wonderful have been, and are being, evolved.

RECOMMENDED
FURTHER READING

DARWIN WROTE ON A PRODIGIOUS VARIETY OF SUBJECTS, from corals to emotions. All of his publications are available through Darwin Online, maintained by John van Wyhe and sponsored by the Arts & Humanities Research Council (UK): http://darwin-online .org.uk.

Additional resources are collected on the companion website to this book: www.themoderndarwin.com.

The body of modern literature directly or indirectly concerned with evolution is dauntingly expansive; happily there are several exceptional books written for a general audience. I highlight some of my personal favorites below – books I find both arresting and thought-provoking – followed by a bibliography.

The foreword recommends Mark Ridley's *How to Read Darwin* for advice on engaging the text of the *Origin*. (Ridley's guidance works as well with this rendition as it does with the original.) Also mentioned are James Costa's meticulously annotated *Origin* and Steve Jones's interpretation in light of modern biology. Additional detailed commentary can be found in David Reznick's *The "Origin" Then and Now: An Interpretive Guide to the "Origin of Species."* These books go some way to demonstrating both the intellectual and scientific environment from which Darwin's work was born and the extent to which that environment has changed. The excellent graphic adaptation by Michael Keller and Nicolle Fuller supplements key excerpts from the *Origin* with explanations and lively illustrations. Among the best accounts of Darwin's life is the brilliant two-volume biography by Janet Browne, also the author of the much shorter but equally fascinating "biography" of the *Origin*.

Ernst Mayr, E. O. Wilson, Stephen Jay Gould, and Richard Dawkins have each written many superb books about evolution (see the bibliography for particular recommendations). Mayr brought his philosophical mode of thought to bear on evolutionary theory, an outlook that helped him rethink our perspective on biological science. Wilson is best known for writing about his lifelong research into sociobiology, while Gould and Dawkins are recognized as passionate public educators.

David Attenborough has become synonymous with the quality nature documentary, and his television programs, including *Life on Earth* and *The Living Planet*, often incorporate evolutionary themes. The companion books to these two series are full of memorable photography and written in Darwin's tradition of wonder at the natural world.

Olivia Judson's *Dr. Tatiana's Sex Advice to All Creation* is the product of scientific expertise crossed with an exceptional facility for engaging writing. The book explores the often incredible consequences of sexual selection. The story of Peter and Rosemary Grant, the husband-wife team that has studied the evolution of finches on the Galápagos for over three decades, is chronicled by Jonathan Weiner in *The Beak of the Finch*.

A basic understanding of molecular biology illuminates a great deal of what was unknown to Darwin, and its concepts are suffused with evolutionary theory. The best-written and most accessible textbook for this grounding is James D. Watson et al., *Molecular Biology of the Gene*.

REFERENCES

Attenborough, David. 1979. *Life on Earth: A Natural History*. Boston: Little, Brown.

———. 1984. *The Living Planet: A Portrait of the Earth*. Boston: Little, Brown.

Browne, Janet. 1996. *Charles Darwin: Voyaging. Volume 1*. London: Princeton University Press.

———. 2003. *Charles Darwin: The Power of Place. Volume 2*. London: Princeton University Press.

———. 2006. *Darwin's Origin of Species: A Biography*. London: Atlantic Books.

Costa, James T. 2009. *The Annotated Origin: A Facsimile of the First Edition of "On the Origin of Species."* Cambridge, Mass.: Belknap Press.

Dawkins, Richard. 2006. *The Selfish Gene (30th Anniversary Edition)*. New York: Oxford University Press.

Gould, Stephen J. 1992. *The Panda's Thumb: More Reflections in Natural History*. New York: W. W. Norton.

Hölldobler, Bert, and Edward O. Wilson. 1990. *The Ants*. Cambridge, Mass.: Belknap Press.

Jones, Steve. 2000. *Darwin's Ghost: A Radical Scientific Updating of the "Origin of Species" for the 21st Century*. New York: Random House.

Judson, Olivia. 2002. *Dr. Tatiana's Sex Advice to All Creation*. New York: Metropolitan Books.

Keller, Michael, and Nicolle R. Fuller. 2009. *Charles Darwin's "On the Origin of Species": A Graphic Adaptation*. New York: Rodale.

Mayr, Ernst. 2001. *What Evolution Is*. New York: Basic Books.

Ridley, Mark. 2005. *How to Read Darwin*. London: Granta.

Reznick, David N. 2011. *The "Origin" Then and Now: An Interpretive Guide to the "Origin of Species."* Princeton, N.J.: Princeton University Press.

Watson, James D., et al. 2013. *Molecular Biology of the Gene*, 7th ed. Cold Spring Harbor, N.Y.: Cold Spring Harbor Laboratory Press.

Weiner, Jonathan. 1995. *The Beak of the Finch: A Story of Evolution in Our Time*. New York: Vintage Books.

INDEX

acclimatization, 87–88

accumulation (of sediment and geological strata), 107, 177, 179, 180, 181, 183, 184, 187, 188, 206, 302. *See also* degradation (of sediment and geological strata)

acquired characteristics, 84

adaptation, 38, 52, 105, 188, 258, 266, 268, 290, 291, 293; of apparently unimportant parts, 121–122; to climate, 87–88; in domesticated organisms, 19, 24, 39, 290; in larvae, 273–275; by natural selection, 67; in social organisms, 54

alligators, fighting among males of, 55

American type, 209, 215, 248

ammonites, sudden extinction of, 197, 199

analogical variation/affinity, xxvi, 31, 32, 120, 123, 258, 297, 301; among related varieties/species, 98–100, 105; versus true affinity, 266–269, 283

ants: and aphids, 131; and neuter castes, 146–149, 285; slave-making instinct of, 135, 136–139

aphids, and ants, 131

armadillo, 90, 208, 210

artificial selection, 67, 122. *See also* selection, human power of

ass: reciprocal cross with horse, 159, 170; striping of, 101–102, 103

auks, 114, 115, 126, 293

Australia, xi, 24, 40, 209; and dogs, 12, 134; and glaciers, 230; as source of florae to New Zealand, 248

Azores, florae of, 224

balance of growth, 91–92, 93

barnacles, x, 62, 94, 119, 179, 189, 264, 274, 278, 279; and balance of growth, 92

barriers to migration, 51, 65, 214–215, 217, 225, 226, 238, 252, 253, 254

bats, 111; as only endemic mammal, xvi, 245, 254, 296; wing of, 93–94, 95, 124, 270, 271, 273, 275, 294, 297

bees, 136, 270, 293; honeycomb-making instinct of, xviii, xix, 106, 139–145; as pollinators, xviii, 46, 58–61

beetles, xix, 38, 48, 53, 240; and use and disuse, 84–85

birds: and birdsong, 55; and colonization of islands, xvi; endemic to the Galápagos, 242, 243; and fear of humans, 132; flightless, xviii, 84, 112, 243, 282; and freshwater mollusk dispersal, 239–240; and inheritance of instinct, 132–135; and seed dispersal, xvii, 2, 39, 223–225, 240–241; structures not in accordance with habit, 114–115, 123–124; and wing structure, 112–113

bizcacha, 215, 267–268

blindness: of cave inhabitants, xii, xxvi, 85–86, 251, 293; in moles, 85, 293; of preconceived opinion, 300

breasts, as rudimentary in male mammals, 280

breeds and breeding, xii–xiii, xvi, 8–13, 19–20, 25–26, 27, 60, 63, 68, 99, 110, 133, 166, 275, 290; and cattle, 147, 198; in confinement, 6, 156, 164; and horses, 219;

DANIEL DUZDEVICH was born in New York and raised in Hungary. Educated at Columbia University and Churchill College, Cambridge, he is currently pursuing a doctorate in biology at Columbia University, studying the interactions between proteins and DNA. Duzdevich was awarded a Paul and Daisy Soros Fellowship for New Americans in 2012.

OLIVIA JUDSON is an evolutionary biologist and award-winning writer based at Imperial College, London. Her first book, *Dr. Tatiana's Sex Advice to All Creation: The Definitive Guide to the Evolutionary Biology of Sex,* has been translated into more than twenty languages including Estonian, Korean, and Turkish; it was also made into a television show. Since then, her writing has appeared in numerous publications, including the *Guardian,* the *Financial Times,* and *National Geographic;* for two years, she wrote a weekly online blog about evolutionary biology for the *New York Times.* She is presently working on her next book.